Delmar Cengage Learning is
proud to support FFA Activities

Join us on the web at
agriculture.delmar.cengage.com

Exploring
AGRISCIENCE
4th Edition

Exploring AGRISCIENCE

Fourth Edition

Ray V. Herren

DELMAR
CENGAGE Learning

Australia • Brazil • Japan • Korea • Mexico • Singapore • Spain • United Kingdom • United States

Exploring Agriscience, 4th Edition
Ray V. Herren

Vice President, Career and Professional Editorial: Dave Garza

Director of Learning Solutions: Matthew Kane

Acquisitions Editor: Benjamin Penner

Managing Editor: Marah Bellegarde

Product Manager: Christina Gifford

Editorial Assistant: Scott Royael

Vice President, Career and Professional Marketing: Jennifer McAvey

Marketing Director: Debbie Yarnell

Marketing Manager: Erin Brennan

Marketing Coordinator: Jonathan Sheehan

Production Director: Carolyn Miller

Production Manager: Andrew Crouth

Content Project Manager: Katie Wachtl

Senior Art Director: David Arsenault

Technology Project Manager: Tom Smith

Production Technology Analyst: Thomas Stover

© 2011, 2006, 2002, 1997 Delmar, Cengage Learning

ALL RIGHTS RESERVED. No part of this work covered by the copyright herein may be reproduced, transmitted, stored, or used in any form or by any means graphic, electronic, or mechanical, including but not limited to photocopying, recording, scanning, digitizing, taping, Web distribution, information networks, or information storage and retrieval systems, except as permitted under Section 107 or 108 of the 1976 United States Copyright Act, without the prior written permission of the publisher.

> For product information and technology assistance, contact us at
> **Cengage Learning Customer & Sales Support, 1-800-354-9706**
> For permission to use material from this text or product,
> submit all requests online at **www.cengage.com/permissions**.
> Further permissions questions can be e-mailed to
> **permissionrequest@cengage.com**

Library of Congress Control Number: 2010921216

ISBN-13: 978-1-4354-3966-5

ISBN-10: 1-4354-3966-X

Delmar
5 Maxwell Drive
Clifton Park, NY 12065-2919
USA

Cengage Learning is a leading provider of customized learning solutions with office locations around the globe, including Singapore, the United Kingdom, Australia, Mexico, Brazil and Japan. Locate your local office at: **international.cengage.com/region**

Cengage Learning products are represented in Canada by Nelson Education, Ltd.

To learn more about Delmar, visit **www.cengage.com/delmar**

Purchase any of our products at your local college store or at our preferred online store **www.CengageBrain.com**

NOTICE TO THE READER

Publisher does not warrant or guarantee any of the products described herein or perform any independent analysis in connection with any of the product information contained herein. Publisher does not assume, and expressly disclaims, any obligation to obtain and include information other than that provided to it by the manufacturer. The reader is expressly warned to consider and adopt all safety precautions that might be indicated by the activities described herein and to avoid all potential hazards. By following the instructions contained herein, the reader willingly assumes all risks in connection with such instructions. The publisher makes no representations or warranties of any kind, including but not limited to, the warranties of fitness for particular purpose or merchantability, nor are any such representations implied with respect to the material set forth herein, and the publisher takes no responsibility with respect to such material. The publisher shall not be liable for any special, consequential, or exemplary damages resulting, in whole or part, from the readers' use of, or reliance upon, this material.

This fourth is dedicated to my wife, Mary Herren who has been a constant source of inspiration and help with all my work. Without her help and encouragement my writing would be much mere difficult and not nearly as much fun.

Contents

Preface / xiii

Acknowledgments / xviii

CHAPTER 1 — The World of Agriculture / 2

Science and Agriculture 6   Growing Plants 8   The Animal Industry 13   Agricultural Exports and Imports 14   Distribution 16   Summary 18   Chapter Review 18

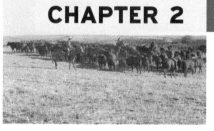

CHAPTER 2 — The History of Agriculture / 22

American Agriculture 25   King Cotton 26   The Cattle Industry 29   Crop Production 31   Scientific Research 35   Changes in Agriculture 37   Summary 38   Chapter Review 39

CHAPTER 3 — Soil: The Origin of Life / 42

Organic Soils 45   Inorganic Soils 46   Water-Deposited Soil 48   Soil Deposited by Wind 49   Soil Texture 50   Soil pH 54   Soil Horizons 55   The Soil Ecosystem 57   Summary 60   Chapter Review 60

vii

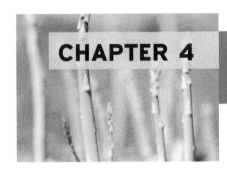

CHAPTER 4 — Plant Structures and Their Uses / 64

Stems 66   Leaves 69   Flowers 71   Seeds 73
Roots 75   Summary 76   Chapter Review 77

CHAPTER 5 — Agricultural Pests / 80

Insects 82   Weeds 88   Genetic Engineering 93
Summary 93   Chapter Review 94

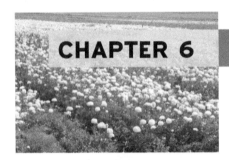

CHAPTER 6 — Floriculture / 98

International Aspect 100   National Aspect 101
Plants for Florists 102   Cut-Flower Production 101
Flower Arrangements 113   Summary 114
Chapter Review 114

CHAPTER 7 — Nursery Production / 118

Growing Media 121   Growing Areas 124   Outside Growing Areas 127   Mist Areas 130   Propagation 131   Plant Production 134   Summary 136
Chapter Review 136

CHAPTER 8 — Landscaping / 140

Design the Landscape 142   Plant Materials 144
Phases of Landscape Development 146
Interiorscaping 161   Summary 163   Chapter Review 164

CHAPTER 9 — Fruit and Nut Production / 168

Pome Fruits 171   Stone Fruits 175   Citrus Fruit 177
Nut Production 180   Summary 183   Chapter Review 184

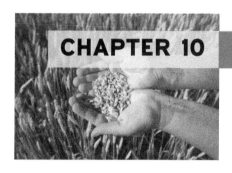

CHAPTER 10 — Row Crops / 188

Grain Crops 190   Oil Crops 197   Fiber Crops 202
Summary 206   Chapter Review 207

CHAPTER 11 — Forest Science / 210

The Natural Forest 214   Forest Succession 215
The Forest Ecosystem 218   Wood-Fiber Production 221   Summary 224   Chapter Review 224

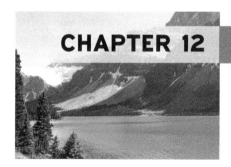

CHAPTER 12 — Protecting the Environment / 228

Pesticides 232   Water Pollution 235   Soil Erosion 239   Nitrate Pollution 242   Wetlands 244
Summary 245   Chapter Review 246

CHAPTER 13 — Organic Agriculture / 250

The Production Process 253   Fertilizers 254
Insect Control 257   Organic Animal Agriculture 258   Summary 262   Chapter Review 262

CHAPTER 14 — The Livestock Industry / 266

The Meat Industry 268   The Horse Industry 280
Small Animals 282   Summary 283   Chapter
Review 284

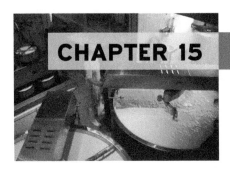

CHAPTER 15 — The Dairy Industry / 288

Milk Production 291   Cheese 296   Yogurt 299
Summary 300   Chapter Review 300

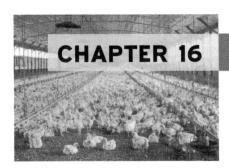

CHAPTER 16 — The Poultry Industry / 304

The Broiler Industry 307   The Layer Industry 315
Turkey Production 316   Other Poultry 317
Summary 319   Chapter Review 319

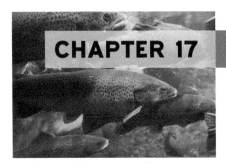

CHAPTER 17 — The Science of Aquaculture / 322

Catfish Production 325   Cool-Water Fish 332
Growing Crustaceans 332   Ornamental Fish 334
Summary 335   Chapter Review 335

CHAPTER 18 — Companion Animals / 335

Health Benefits 342   Service Animals 345   Diseases
and Afflictions 348   Industry 350   Health Care 351
Responsible Ownership 352   Summary 354
Chapter Review 354

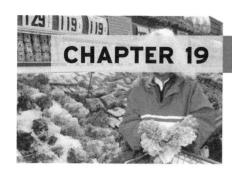

CHAPTER 19 Preserving Our Food Supply / 358

Method of Perserving Foods 361   Food Safety 368
Summary 371   Chapter Review 372

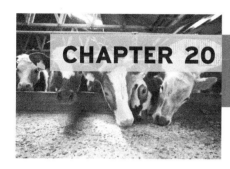

CHAPTER 20 The Ethical Treatment of Animals / 376

Animal Welfare 379   Animal Rights Activists 387
Summary 389   Chapter Review 389

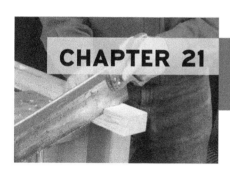

CHAPTER 21 Selecting and Using Hand Tools / 394

Introduction 396   Simple Machines 397   Working With Hard Woodworking Tools 399   Measuring and Marking Tools 401   Summary 415   Chapter Review 416

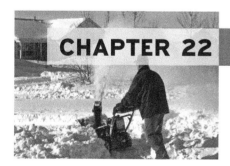

CHAPTER 22 Small Engine Operation / 420

Small Engine Safety 423   Operating Principles of small Engines 424   The Four-Stroke Cycle Engine 428   The Two-Stroke Cycle Engine 431   Comparing Four- and Two-Stroke Cycle Engines 433   Diesel Engines 434   Engine Systems 435   Summary 446   Chapter Review 446

CHAPTER 23 Biofuels / 444

Ethanol 448 Biodiesel 452 Biomass 456
Methane 458 Concerns 459 Summary 460
Chapter Review 461

CHAPTER 24 Biotechnology: The Future of Agriculture / 464

Biotechnology 466 Careers in Agriscience 474
Summary 475 Chapter Review 476

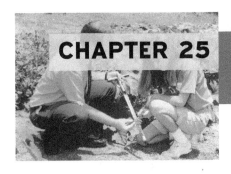

CHAPTER 25 High School Agricultural Education Programs / 480

Modern Programs in Agriculture 482
Summary 490 Chapter Review 491

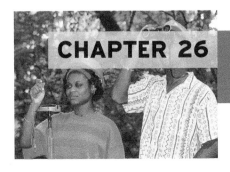

CHAPTER 26 Careers in Agricultural Science / 494

Plant Science 496 Animal Science 500 Food
Science 503 Natural Resources 507
Summary 509 Chapter Review 509

Glossary / 512

Index / 524

Preface

Agriculture is a rapidly changing science. Change comes about so rapidly that it is difficult to keep up with all the developments. However, basic scientific principles are the foundation of modern agriculture and these principles drive change. *Exploring Agriscience* Fourth Edition is meant to be the first text that introduces students to the dynamic industry of agriculture. Integrated throughout are principles of science behind the industry. Research has shown that most Americans have a misconception of agriculture. Many think of agriculture as farming. While this is a basic component of agriculture, the scope of agriculture is much broader. This industry employs more people than any other and is responsible for much of the nation's wealth. This text provides students with an overview of the different aspects of that industry. Of course, all aspects of agriculture are not covered. To address all phases of the industry would take many volumes and would be beyond the confines of classroom teaching. Rather, this text introduces the main areas of agriculture that have made the industry great.

The fourth edition of *Exploring Agriscience* contains one new chapter along with many updates in the previous chapters. The new chapter on biofuels addresses a topic that is a current ongoing issue. Students need to be aware of the finite nature of fossil fuels and the potential of alternative fuels. In the fourth edition, a majority of the images have been updated. Over 250 new images have been added to replace dated images. This should be a great help in teaching students the modern concepts of the industry of agriculture and agriscience.

All aspects of the agricultural industry are based on science. Principles of biology and physics have been researched, manipulated, and used to provide advancements in the production of food and fiber. These advancements have made life better for all Americans. No education would be complete without an understanding of the role of agriculture in our lives.

This book was written with the hope that students will continue the study of the science of agriculture. Within each chapter are highlights of particular FFA activities for high school students. These are by no means the only programs of the organization. They represent areas that coincide with the content area of each chapter. Any student with an interest in biological science will find a program in Agriculture Education and the FFA that will match that interest.

The addition of the chapter on high school agriculture programs will provide students with more insights into the program.

This revised title from Delmar Cengage Learning is written to be used as a core text for a middle school agriscience curriculum.

Features

- Full color photos, illustrations, and design stimulate student interest and make difficult concepts easier for student comprehension.

- Boxed articles entitled "Career Development Events" feature FFA activities and competitions.
- Numerous hands-on activities and discussion questions reinforce learning of material and concepts covered in the text.

Enhanced Content

- Provides a hands-on exploratory, science-based approach to agriscience designed to be used as a student's first text in agriculture.
- Broad-based coverage of horticulture, animal science, environmental science, and biotechnology is presented in an easy-to-understand format for middle school students.
- A new chapters, Biofuels, reflects Delmar Cengage Learning's commitment to continue bringing the latest trends in agriculture education to the classroom.

Extensive Teaching/Learning Package

The complete supplement package was developed to achieve two goals:

1. To assist students in learning the essential information needed to continue their exploration into the exciting field of agriscience.
2. To assist instructors in planning and implementing their instructional program for the most efficient use of time and other resources.

Instructor's Guide to Text

The Instructor's Guide provides answers to the end-of-chapter questions and additional material to assist the instructor in the preparation of lesson plans.

Lab Manual

This comprehensive lab manual reinforces the text content. It is recommended that students complete each lab to confirm understanding of essential science content. Great care has been taken to provide instructors with low cost, strong content labs.

Lab Manual CD-ROM

Our comprehensive student lab manual is available on a CD-ROM. Instructors have the capability of printing off as many copies of the exercises as needed, for as long as they own the book.

Lab Manual Instructor's Guide

The Instructor's Guide provides answers to lab manual exercises and additional guidance for the instructor.

Classmaster CD-ROM

This supplement provides the instructor with valuable resources to simplify the planning and implementation of the instructional program. It includes a pdf of the instructor's guide, a computerized test bank, instructor slides in PowerPoint and an image library to provide the instructor with a robust set of tools for presenting each topic.

Classroom Interactivity CD-ROM

This supplement provides instructor's the opportunity to create a dynamic learning environment while engaging their students in active participation. This CD-ROM contains four different "game-show-themed" applications to be run by the instructor. All questions are taken directly from the readings of the text and serve as great tools in reinforcing the main concepts of each unit.

Acknowledgments

The author wishes to thank the following for their help in preparing the fourth edition: Jenna Brown for her help in composing materials and helping with the Instructor's Guide; Arenthia Herren for her help in writing the chapter on Biofuels and Dr. Frank Flanders for his help in locating images for the text.

The author and Delmar would like to thank the reviewers who shared their content expertise and added valuable suggestions for this fourth edition:

Maxine Chipman
Wonewoc-Center Schools
Wonewoc, WI

Jason Blair
Madison High School
Rexburg, ID

Art Design Credits

Cover

Test tubes: © Sandra Cunningham/Shutterstock
Wind Turbine: © Nikada/iStockphoto
Horses: © Julie Johnson/iStockphoto
Tractor: © PhotoDisc
Flowers: © PhotoDisc
Field: © PhotoDisc

Design Art

Chapter Openers
© Sandra Cunningham/Shutterstock

Preface
© Mike Bentley/iStockphoto

Acknowledgements
© Rob Friedman/iStockphoto

Table of Contents
© Adrian Beesley/iStockphoto
Chapter 1: © Sandra Cunningham/Shutterstock
Chapter 2: © Lenice Harms/Shutterstock
Chapter 3: © Adrian Beesley/iStockphoto.
Chapter 4: © Duöan Zidar/iStockphoto

Chapter 5: © iStockphoto
Chapter 6: © Eldad Yitzhak/iStockphoto
Chapter 7: © slowfish/Shutterstock
Chapter 8: © iStockphoto
Chapter 9: © Daniel Mar/iStockphoto
Chapter 10: © Bart Sadowski/iStockphoto
Chapter 11: © iStockphoto/Finn O'Hara
Chapter 12: © iStockphoto
Chapter 13: © Hedda Gjerpen/iStockphoto
Chapter 14: Courtesy of Dr. Frank Flanders
Chapter 15: © Kheng Guan Toh/Shutterstock
Chapter 16: Courtesy of USDA
Chapter 17: © Achim Prill/iStockphoto
Chapter 18: Delmar/Cengage Learning
Chapter 19: Courtesy of USDA
Chapter 20: © David Cole/iStockphoto
Chapter 21: © Gillian Mowbray/iStockphoto
Chapter 22: © Joe McDaniel/iStockphoto
Chapter 23: © Logan Buell/iStockphoto
Chapter 24: © David R. Gomez/iStockphoto
Chapter 25: Courtesy of Dr. Frank Flanders
Chapter 26: Courtesy of USDA

CHAPTER 1
The World of Agriculture

Student Objectives

When you have finished studying this chapter, you should be able to:

- Discuss why agriculture is important in our everyday lives.
- Explain why agriculture is such a broad and diverse industry.
- Explain why agriculture is a science.
- Summarize the reasons why agriculture contributes to the wealth of the nation.
- List and discuss the uses of plants.
- List and discuss the uses of animals.
- Contrast agricultural exports and imports.
- Discuss the processing and distribution of agricultural products.

Key Terms

agriculture
biology
ecology
export

import
commodities
growing season

You are about to begin a study of one of the most dynamic industries the world has ever known! This industry—American agriculture—is involved with growing plants and animals for use by all the people of this country and much of the world. We live in a country with one of the highest standards of living in the world. Americans are the envy of most of the world! Much of the affluence we enjoy is either a direct or indirect result of the agriculture industry. Compared to people in other countries, we pay a very small portion of our income for food. We spend around 10 percent of our income on food (Figure 1–1). This compares to over 70 percent in some countries. Our prosperity began because of the fertile soil and ideal climate we enjoy in the United States. Our soil and climate provided the basis for our great agricultural industry. New discoveries and developments as a result of scientific research in agriculture promoted the growth of the agriculture industry. This industry has been one of the major causes of the wealth of the country.

The American Farm Bureau has compiled the following list about the importance of American agriculture:

- Agriculture is America's leading export.
- About 17 percent of raw U.S. agriculture products are exported yearly.

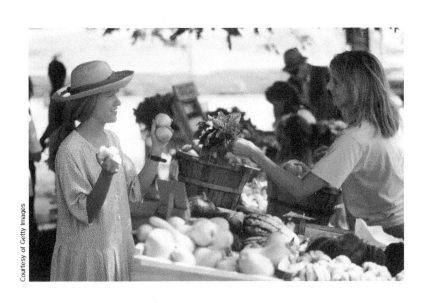

FIGURE 1–1

Americans only spend about 10 percent of their income for food.

THE WORLD OF AGRICULTURE

- One-fourth of the world's beef and nearly one-fifth of the world's grain, milk, and eggs are produced in the United States.
- One in three U.S. farm acres is planted for export, and 25 percent of gross farm income comes directly from exports.
- Through research and changes in production practices, today's food producers are providing Americans with the widest variety of foods ever.
- Research and advancements in biotechnology are now in the marketplace, with tastier fruits and vegetables that stay fresh longer and are not damaged by insects.
- Consumers derive health benefits from changes in farm production, including less fat in meat. Tofu, a soybean product, has been shown to reduce the risk of some cancer and heart disease.
- With modern methods, 1 acre of land in the United States (about the size of a football field) can produce: 42,000 lbs. of strawberries, 11,000 heads of lettuce, 25,400 lbs. of potatoes, 8,900 lbs. of sweet corn, or 640 lbs. of cotton lint.
- American consumers spend the lowest percentage of their annual income on food—around 10 percent.
- Cotton is by the far the most dominant fiber produced in the United States and is used for apparel and home fabrics, as well as industrial uses.
- The agricultural industry employs 20 percent of America's workforce, or approximately 21 million people.
- Farmers and ranchers are the first environmentalists, maintaining and improving the soil and natural resources to pass on to future generations.

- Farmers use reduced tillage practices on more than 72 million acres to prevent erosion.
- Farmers maintain over 1.3 million acres of grass waterways, allowing water to flow naturally from crops without eroding soil.
- Contour farming, planting crops on hillsides instead of up and down, keeps soil from washing away. About 26 million acres in the United States are managed this way.
- Cattle ranchers and others control water run-off with sod waterways and diversions, erosion control structures, and catch basins.
- Just as urban families recycle grass, newspaper, and aluminum, farm families have practiced recycling for a long time by applying manure to fields to replace nutrients in the soil.
- Agricultural land provides habitat for 75 percent of the nation's wildlife.

Science and Agriculture

Science is the study of or the explanation of natural phenomena. In other words, science deals with the laws of nature and our understanding of them. Agriculture fits this definition very well. No other discipline deals more directly with the laws of nature than does agriculture. Recognized areas of science such as physics, chemistry, geology, meteorology, and biology all are a part of agriculture. Biology is the study of living organisms—their life cycles and how they live, grow, and reproduce. All of the knowledge we have of living things (biology) has only three applications: medicine, ecology, and agriculture. In fact, medicine and ecology (the study of the relationship of organisms in an environment) are related to agriculture. Medicine is related to agriculture because good nutrition is dependent on humans

FIGURE 1-2
One American farmer feeds more than 100 people.

obtaining plenty of wholesome food. Ecology is related because of the tremendous impact agriculture has on the environment.

Agriculture is the most important of all the sciences because we depend on agriculture for the food we eat. Without a steady intake of food we would soon die. Through the scientific developments that have made agriculture so efficient, fewer than half of the people in this country are involved with getting food to the rest of the population. Our system has become so efficient that one American farmer feeds more than 100 people (Figure 1-2). Of course, a large number of people support the farmer, in providing supplies and distributing the produce. This allows all of the people not involved in agriculture to produce manufactured goods and conduct services that would otherwise be impossible. If it were not for our agricultural system, people would spend most of their time gathering something to eat.

Many people think of agriculture in terms of a farmer out plowing a field, and this is a part of agriculture. However, farmers represent a small percentage of the people who work in agriculture. The agriculture industry hires more people each year than any other industry (Figure 1-3). These jobs involve such diverse occupations as marketing

FIGURE 1-3
The industry of agriculture hires more people than any other industry.

specialists, chemical manufacturers, food inspectors, and research scientists. Yet all of these people are a part of agriculture.

The growing of plants and animals involves many different segments. At one time, the growing of living things was only for food and clothing. In modern times in this country, many other uses have been made of the plants and animals we grow. In fact, new uses are constantly being found. For example, for many years our country has relied on other countries to supply a large portion of our petroleum that is used for fuel. We are now beginning to grow plants that supply fuel to meet our energy needs. Corn is used to produce ethanol for fuel to replace gasoline. Diesel engines were originally developed to run on vegetable oil, and modern diesel engines can be made to run quite well on vegetable oils such as peanut or soybean oil.

Growing Plants

In a very real sense, the foundation of all agriculture is the growing of plants. Without plants, the animals we grow for food would have nothing to eat. Much of the food we eat comes from plants (Figure 1–4). The flour that goes into our breads and the oil that we use for cooking come from plants. The fresh fruits that supply us with vitamins and other nutrients are

FIGURE 1–4

Agriculture supplies us with nutritious food.

FIGURE 1-5
The growing of trees is a part of agriculture.

plant products. Just think of all the vegetables you ate during the past week! All of these were produced by our agriculture industry. The seeds of plants supply us with a rich source of protein. Did you know that peanut butter is made from the roasted and ground-up seeds of a plant?

Plants also supply us with clothing. The fiber from cotton plants gives us comfortable clothing. This crop has been grown in this country since before it was a country. Today, only China grows more cotton than the United States.

Our houses also come from plants. Trees supply the lumber for buildings and furniture. You might not have thought of forestry as being agriculture, but most of the trees that are harvested each year have been planted and cared for by tree producers (Figure 1-5). Trees also supplied the paper to make this book. Just think of all the paper Americans use every day. This industry represents a large portion of the economy of the nation.

Many medicines come from plants. A good example is the drug digitalis that is used to help people with heart problems. This important medicine comes from the foxglove plant, which is grown by agricultural producers (Figure 1-6). Currently over 120 different chemicals that come from plants are used as important ingredients in commercially produced drugs.

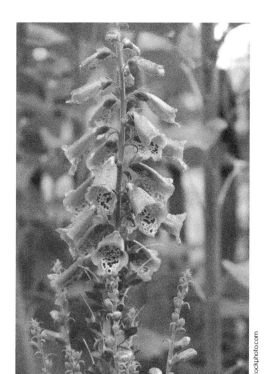

FIGURE 1-6
Plants like this foxglove plant provide us with medicines.

One large and expanding aspect of agriculture is the growing of ornamental plants. Americans spend millions of dollars each year buying plants to beautify their homes. Drive through almost any neighborhood and you will see a broad variety of plants used to enhance the value of homes. Trees, shrubs, grass seed, and turf are all grown on farms and nurseries. Potted plants and flowers are

CAREER DEVELOPMENT EVENTS

When you enroll in high school Agriculture Education courses, you become a member of the FFA Organization. This organization gives you opportunities you may have never even dreamed possible. The skills and concepts you learn in the classroom and FFA activities are a fun way to learn. You can even travel to other countries to learn about agriculture.

The first FFA International Exchange Program was with England in 1948. Since then the FFA international program has grown along with the realization that a world food supply is an important part of agriculture in the United States. FFA international activities are designed to link FFA members with other nations to provide experiences that will improve agriculture education around the globe. There are several ways in which an FFA member can become involved in an international exchange program.

In your local FFA school chapter, you can be part of the international program by promoting and conducting FFA development projects. The development projects can involve single FFA members and chapters nationwide. The FFA can help by improving agriculture practices and crops in developing countries, helping world hunger problems, establishing rural agriculture education programs, and developing new technology in agriculture. Projects for FFA chapters range from raising funds to support projects to purchasing books, tools, or seed for a rural agriculture education facility.

The chapter may even sponsor and house an exchange student from another country. The World Agriscience Studies (WASS) is an international agricultural student exchange program. Students from other nations are placed with American host families to study agriculture in the United States. The local FFA produced in greenhouses for people to buy. Each year, many people make their living raising plants in greenhouses.

Plants are used in ways you probably never even considered. Did you realize that chewing gum comes almost entirely from plants? The gum is made from a substance called chicle that comes from the sap of a tree. The sugar used to sweeten it comes from sugar beets or sugarcane (Figure 1–7). Even flavorings such

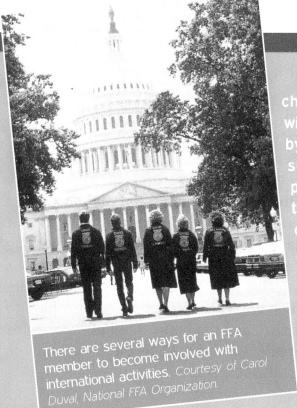

There are several ways for an FFA member to become involved with international activities. *Courtesy of Carol Duval, National FFA Organization.*

chapter can provide foreign students with experiences in agriculture by involving them in their local supervised agriculture experience projects. FFA members also have the opportunity to travel to a foreign country, live with a host family, and attend a year of high school, studying an agricultural curriculum.

Other ways to become involved with international agriculture are through study seminars and work experience programs. Agricultural study seminars are organized annually for state associations. The seminars last two to three weeks and usually include much traveling. The seminars typically include visits and tours through agribusiness industries, research labs, agriculture schools, and a wide diversity of farms. Work Experience Abroad (WEA) is an international experience program for FFA members in college or alumni FFA members who work in the agriculture industry. FFA members are placed in other nations to work on farms or in agribusiness for 3 to 12 months. The program also coordinates the hosting of students from other nations who will come to the United States. The experience of hosting an exchange student can be as meaningful as traveling abroad.

FIGURE 1-7
The sugar we use comes from crops such as sugarcane.

as mint come from cultivated plants. All these plants are planted, cultivated, and harvested to produce chewing gum. This makes chewing gum an agricultural product!

The growing of the plants is supported by other parts of agriculture. Fertilizers are produced that supply the plants with nutrients. Chemicals are manufactured to produce pesticides that kill weeds and insects that attack the plants. Tractors, cultivators, and harvesting equipment have to be designed and built (Figure 1-8). Fuel and lubricants have to be produced for the tractors and equipment. Large irrigation systems have to be built, installed, and maintained. All of these items have to be marketed and sold to the producer. The growing of plants is big business!

FIGURE 1-8
All of the machinery used in agriculture has to be designed and built.

The Animal Industry

Most of the animals produced in this country are raised for food. The cattle industry supplies us with beef. The steaks you eat for dinner and the hamburgers you buy at the fast food restaurant come from the cattle industry (Figure 1–9). The swine industry gives us such meat products as pork chops, ham, and sausage. The sheep industry supplies meat for a large part of our population. The poultry industry produces meat for such products as fried chicken and chicken potpies. The poultry industry provides a source of nutritious, inexpensive meat and protein-rich eggs. Fish, shellfish, and other aquatic animals are grown by producers to be used as food. The dairy industry supplies us with fresh milk, cheese, butter, cream, yogurt, ice cream, and other products.

The animals that are slaughtered give us other products. Leather for making shoes, belts, and other clothing comes from cattle, hogs, and other animals. In addition, several medicines and other by-products are made from animal products.

Animals are still used for work. In many places in the western part of our country, horses are used to herd cattle. They are used as pack animals. Horses provide us with recreation through riding and horse racing (Figure 1–10). Dogs are also produced for

FIGURE 1–9
Steak for our dinners comes from the cattle industry.

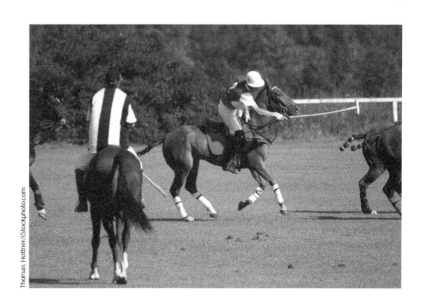

FIGURE 1–10
Horses provide us with recreation.

work. They are valuable in herding sheep and are used in moving cattle and hogs.

An increasing part of the animal industry is the raising of companion animals. Dogs are raised and trained to be used to help visually impaired people, and are raised for pets. The pet industry is broad and diverse. Animals from goldfish to cats are raised as companions to humans.

Like the plant industry, the animal industry has a lot of support areas. Animals have to be fed. The feed industry supplies producers with high-quality feeds that contain the nutrients animals need (Figure 1–11). Medicines are needed to keep the animals healthy. Pesticides are needed to keep the animals free from parasites. Fencing has to be produced. Buildings and facilities have to be built to house the animals. Trailers and trucks are needed to transport the animals. Buyers, brokers, inspectors, and processors are all involved in the industry. And it is all a part of agriculture.

FIGURE 1-11
The feed industry provides feed for livestock.

Agricultural Exports and Imports

An export is a product that is shipped from the United States to a foreign country. An import is a product that is brought into this country from another country. Agriculture in this country both exports and imports commodities (Figure 1–12). Commodities are any useful things that can be produced, sold, or bought. Commodities are the end result of agricultural production. The trade balance of agricultural products is good news for our country. The trade balance refers to the value of goods shipped out of this country (exports) compared to the value of goods shipped into this country (imports). In recent years, some industries have lost a large share of their market to companies in other countries. However, American agriculture still dominates the world's agricultural demand. Our wheat, corn, soybeans, and other products are shipped all over the world to feed people. Every

THE WORLD OF AGRICULTURE

FIGURE 1-12
Agriculture in this country imports and exports goods. This ship is loaded with wheat.

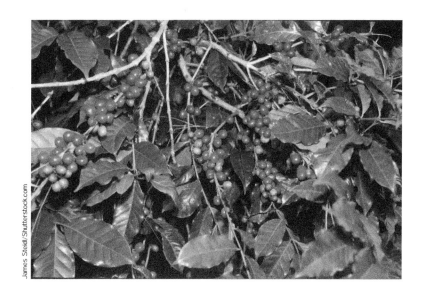

FIGURE 1-13
Coffee is a crop that does not grow well in the united states.

year we export almost 70 percent more agricultural goods than we import. Each day we export around $6 million in agricultural products to other countries around the world. This generates more than $100 billion annually for our economy by creating business activities and millions of jobs.

Some agricultural products are brought into this country. Although we have a near ideal climate for agricultural production, some crops grow better elsewhere. For example, coffee grows best in the mountainous areas of South America (Figure 1-13). Bananas grow in the tropical areas of Central America, and many specialty foods such as cocoa and vanilla are brought in from other countries.

Another reason for importing agricultural products is so that we can enjoy fresh produce year-round. Because much of South America is south of the equator, their seasons are opposite from ours. This means that when it is winter here it is summer there. We take advantage of this by bringing such produce as grapes into the United States so we can enjoy them all year-round. Countries such as Mexico that are closer to the equator have a longer growing season and can produce vegetables and melons earlier than producers in the United States. By shipping them in from Mexico, we can enjoy watermelons even before it is time to plant them in most parts of this country.

All of the industries involved in exporting and importing agricultural products create a lot of jobs. This adds to the economy and well-being of our nation. When the imports of manufactured products exceed the exports, agricultural exports help to make up the gap.

Distribution

FIGURE 1-14
Agriculture products are shipped from one part of the country to another.

If you are an average person you probably eat beef from Iowa, apples from Washington State, cheese from Wisconsin, tomatoes from California, potatoes from Idaho, and oranges from Florida. You wear clothes made of cotton from Arizona and live in a house built from lumber grown in Oregon. Of course, other states also produce these products—agricultural products come from a wide area. The states mentioned are ideal for the production of these items because of their climate and other factors. Cotton grows poorly in Washington State, and Arizona is more suited to growing cotton than timber.

Agricultural products are grown where the climate and other factors are suitable for production. However, people in Arizona need lumber, and people in Washington State need cotton. To solve this problem, agriculture has a marvelous system for distributing products all over the country (Figure 1-14). Many people make their living buying farm produce from

THE WORLD OF AGRICULTURE

producers and selling to wholesale distributors or to processors. These people process and package the produce and sell to retail stores that offer the produce to the public. All this involves a complicated network of transportation. The produce is carried by railroad or by truck. Produce that is easily spoiled may be transported in a refrigerated railcar or truck.

Food and fiber processing is a huge industry that is a part of agriculture. Consumers want products that are as nearly ready to eat as possible (Figure 1–15). People no longer want to wash, shell, and cook beans in the hulls as they come from the producer. Instead, they prefer to have the beans processed and frozen in packages that go into the microwave oven and come out ready to eat. Few people want to make a pizza from scratch when they can have one that is ready to be popped into the oven for a few minutes. Think a minute about all of the processing that had to go into making that pizza. The tomato sauce had to be made from raw tomatoes and spices. The crust was made from flour that was processed from wheat. The cheese had to be processed from milk, and the sausages had to be processed from various meats and spices. All of the processing required that large numbers of people work together just to get the pizza ready for you—and they are all a part of agriculture.

FIGURE 1–15
Americans prefer to buy their food ready, or almost ready, to eat.

SUMMARY

Look around you. Agriculture is very much a part of your life. Without it our lives would be dramatically changed. In the following chapters we will explore the wonderful world of American agriculture.

CHAPTER REVIEW

Student Learning Activities

1. Make a list of all the jobs in your community that are involved in agriculture. Choose a particular job and list all the qualifications one would need for the job.
2. Go to the library and research the agricultural products grown in your state. Write an essay on these products. Which are the most important? Are the products used within the state, or are they shipped to other places?
3. Interview 10 people about their perceptions of agriculture. Draw a conclusion as to whether or not the majority of the perceptions are correct.
4. Keep a journal of all your activities for a week. At the end of the week, list all the ways agriculture has affected your life.

True/False

1. _____ Physics, chemistry, geology, meteorology, and biology are all part of agriculture.
2. _____ The foundation of all agriculture is growing plants.
3. _____ Most of the animals produced in this country are raised for producing dairy products.
4. _____ U.S. agricultural imports greatly exceed exports.
5. _____ Compared to people in other countries, we pay a very large portion of our income for food.
6. _____ Soil and climate provide the basis for the agricultural industry.
7. _____ A large portion of the medicines we use comes from plants.
8. _____ No other country produces more agricultural products than the United States.

9. _____ Companion animals assist farmers with work on the farm.

10. _____ One large and expanding aspect of agriculture is the growing of ornamental plants.

Multiple Choice

1. The study of or the explanation of natural phenomena is called
    a. agriculture
    b. science
    c. farming

2. One American farmer feeds more than
    a. 100 people
    b. 200 people
    c. 500 people

3. The country that leads in the production of cotton is
    a. the United States
    b. India
    c. China

4. Most of the animals produced in this country are raised for
    a. work
    b. dairy products
    c. food

5. No other discipline deals more directly with the laws of nature than does
    a. agriculture
    b. horticulture
    c. floriculture

6. Agriculture is the most important of all the sciences because we depend on it for
    a. dairy products
    b. jobs
    c. food

7. The foundation of all agriculture is
    a. raising animals
    b. growing plants
    c. growing fruit

8. The seeds of plants supply us with a rich source of
    a. carbohydrates
    b. protein
    c. vitamins

9. Each year many people make their living raising plants in
   a. greenhouses
   b. their homes
   c. gardens

10. Pesticides are needed to keep animals free from
    a. predators
    b. parasites
    c. poor weather conditions

Discussion

1. How are medicine and ecology related to agriculture?
2. List some jobs involved with agriculture.
3. Why are some plants so important in our daily lives?
4. What are companion animals?
5. What is needed to raise healthy animals?
6. List the uses we have for animals.
7. Why is agriculture the most important of all sciences?
8. Where does chewing gum come from?
9. Why is food and fiber processing such a large part of agriculture?
10. List the three applications available for all of our knowledge about living things.

CHAPTER 2
The History of Agriculture

Student Objectives

When you have finished studying this chapter, you should be able to:

- Describe the importance of agriculture.
- Discuss how agriculture was necessary to the development of civilizations.
- Tell how the cotton industry brought wealth and created problems in the United States.
- Describe how the cattle industry developed.
- List some of the inventions that had an impact on agriculture.
- Discuss how scientific research helped develop agriculture.
- Define the Land Grant concept of education.
- Explain how the Land Grant institutions have helped shape agriculture.

Key Terms

civilizations
hunters and gatherers
scientific research
Land Grant Act
USDA
experiment stations

The science of agriculture is almost as old as the human race. In the history of humans on the planet, many civilizations have come and gone. Some were powerful and warlike and some were small and peaceful. Some civilizations, such as the Greeks and Romans, provided the foundation for future civilizations. Architecture, art, literature, and political theories came from these civilizations. Other peoples accomplished great feats of engineering and building. Many aspects of ancient civilizations were different, but they all had one thing in common. Before any civilization could exist and flourish, a strong agricultural base had to be established (Figure 2–1). The reason for this was simple: Before roads could be built, buildings designed, or works of art created, the people had to be fed.

Before the creation of agriculture, people had to spend most of their time hunting and gathering food. These people are known as hunters and gatherers. In order to survive, they had to find and collect the food nature had to offer. This was a full-time job for all people, and little time could be spent doing anything else. This caused the people to have to travel all the time because food in an area would soon be depleted. When people found they could grow their own food, many of their problems were solved. Now they have

FIGURE 2–1

Before any civilization can flourish it must have a strong agriculture.

FIGURE 2-2
When people could grow their own food, they could stop wandering and build villages.

a ready supply of food, and could now settle down in one place (Figure 2-2).

Shortly after people began to plant seed and tame animals, they began to search for better ways of growing food. By trial and error they discovered the best time to plant, the best plants to grow, and the best animals to raise. As they got better at growing plants and raising animals, more food could be produced. As more and more food was produced, it took fewer people to grow the crops and keep the animals. In time, only certain people grew food, and others were then free to accomplish other things. A system of trade was established where goods such as pottery or clothing were traded for the food produced by the farmers. When farmers became really proficient, people could settle in one place and create villages. They could then build roads and towns and create art. This was because other people produced the food needed to feed the workers and builders. The stronger the agriculture of a civilization, the stronger could be the army and groups of workers. Without plenty of food to feed all of the people, little progress can be made by any people.

American Agriculture

The United States is among the wealthiest nations on the earth. Much of this wealth can be attributed to the tremendously successful agriculture system.

FIGURE 2-3

American agriculture is the envy of the world.

As pointed out in the first chapter, American agriculture is the envy of the world (Figure 2–3). No one even comes close to matching the agricultural production of this country.

Our agricultural system began before we were a country. Most of the immigrants who first came to the New World were farmers. The colonists made their living producing crops that could be sent back to Europe. At the time of the American Revolution, well over 90 percent of the colonists made their living through agriculture. In fact, most of the signers of the Declaration of Independence were farmers. At this time the main crops grown were tobacco, sugarcane, and rice. The produce that was not used by the colonists was sent to England, where the products were exchanged for manufactured goods. The rich soil and climate made the New World an ideal place for agriculture. These conditions allowed the development of a variety of crops. But before agriculture could advance very far, certain developments had to take place. The following is a description of some of these developments.

 ## King Cotton

Perhaps the first huge impact that agriculture had on the wealth of the nation came about as a result of cotton production. Cotton had been grown in India,

THE HISTORY OF AGRICULTURE

FIGURE 2-4
Upland cotton has seeds that seem to be glued to the lint.

Egypt, and China for thousands of years. It had also been grown in the New World in what is now known as Peru and Mexico from 2500 B.C. It was introduced into the coastal areas of Georgia and South Carolina in very early colonial days. The fiber from the plant was much prized as a material to use in making clothes. It was easy to spin and weave, easy to dye, and comfortable to wear.

There were two types of cotton—upland and Sea Island cotton. Sea Island cotton was the most widely grown. The fibers were longer (called staple length), and the seeds could be easily removed from the fibers. The problem with upland cotton was that the fibers seemed to be glued to the seeds and were extremely difficult to remove (Figure 2-4). Even though it was a slow process, the seeds from the Sea Island cotton separated from the seeds by hand very easily. The problem was that this type of cotton would only grow along the coasts of the southern colonies.

About 1790, an inventor named Eli Whitney built a machine called a cotton engine (the name was later shortened to cotton gin) that could remove the seeds from upland cotton (Figure 2-5). The device consisted of spikes on a revolving drum that passed through slots and separated the lint from the seeds. A rotating brush then removed the lint from

FIGURE 2-5
The cotton gin removed seeds from the cotton lint.

the seeds. One person turning the crank of the gin could remove more seeds than many dozen people removing them by hand. When the gin was fully developed and powered by steam, removing seeds was no longer an obstacle to growing cotton. This new device had a profound effect on the history of our country. For the first time, upland cotton could be profitably grown. This meant that a large area from what is now southern Missouri to Virginia and all of the areas south of there could be used to produce cotton. Cotton was an ideal crop to grow. Only so much tobacco, sugar, and rice could be used and the market could be easily flooded. But cotton was a product that everyone could use. Now that the seeds could be efficiently removed, a gigantic market opened for cotton. As the demand grew, more cotton was produced. When this happened, the price fell. This caused the cloth to be affordable to poor people as well as rich. This type of clothing was cheaper than the traditional wool or linen. It also had characteristics such as comfort that people wanted. When the price fell, more people could buy it and demand increased.

As more and more cotton was produced, new fields to the west in Alabama, Tennessee, and Mississippi were cleared and planted. A family could clear the ground and plant, cultivate, and harvest enough cotton to make a living (Figure 2-6). Although a lot of cotton was produced by large plantations using slave labor, even more cotton was produced by families who worked the land themselves. The entire southern portion of the United States was settled primarily because of the cotton industry. Cotton became so important that it became known as King Cotton.

Most of the cotton produced went to Europe (Figure 2-7). The textile factories of England, France, and Germany ran night and day using cotton from the United States. Ships that delivered the cotton returned with manufactured goods that were sold to the people of the southern states. By the 1850s these industries were almost totally dependent on American cotton.

FIGURE 2-6
One family could clear the land and make a living producing cotton.

FIGURE 2-7
In the mid-1800s most of the cotton produced in the United States went to Europe.

While the South was developing its economy on cotton, the northern states were developing industries. These two diverse sections of the country developed disputes over the role of the federal government, taxation, and about the issue of slavery. Ultimately, these differences led to the American Civil War. The development of agriculture had had a profound effect on the young country!

The Cattle Industry

The cattle industry in this country had a slow beginning. Beef was expensive. In the areas close to population centers, cattle had to be fed expensive grain or had to be fed on pastures that could be put to better use growing crops. In the newly settled areas of the West, cattle could be raised on vast areas of grass (Figure 2-8). These areas were suited to the growing of cattle but not crops. The only problem was that the cattle-grazing areas were far from the centers of population.

After the American Civil War, western cattle producers decided the best way to get their cattle to the large cities was to ship cattle on the new railroads

FIGURE 2-8
In the newly settled territories of the west, cattle could graze on the vast areas of grass.

FIGURE 2-9
Cattle were driven to railheads where they were shipped to the East.

that had recently stretched westward. This would provide a way of getting the live cattle close enough to the consumer (Figure 2-9). This was necessary because the animals had to be slaughtered and consumed within a short period or the meat would spoil. The big problem was that there were only a few places where the railroad ran. Certain areas such as central and western Texas had no rails at all. Towns like Sedalia, Missouri, and Abilene, Kansas, became established as "railheads" where cattle were brought for shipment. The only way to get the cattle to the railheads was to drive them. The fabled cattle drives of the West were started in this manner. For a period of about 20 years, this was the means of getting cattle to market. This allowed cattle producers to make a profit by using the millions of acres of grassland to graze cattle.

After the railroads were finally completed around the turn of the twentieth century, cattle could be delivered live without the long arduous drives. A problem still remained. Even the rail journey took several days. During this time, the cattle lost weight from stress of the long trip. Producers reasoned that more profit could be made if the cattle were slaughtered close to where they were raised and the meat shipped. However, this was impossible because the meat would spoil before it could reach market.

People had known for centuries that if meat was kept cool it could be kept longer without spoiling. The first attempt at cooling meat was the use of ice that had been cut from frozen lakes during the winter and stored in icehouses. The ice blocks were suspended from the ceiling in the meat storage rooms in an effort to keep the meat cool. This effort was not very successful. The ice melted quickly and did not keep the meat cool enough. Then, during the 1880s, mechanical refrigeration was developed. This allowed storage rooms to be kept cool all year long.

A few years later the refrigerated boxcar was invented. This innovation allowed the transportation of meat anywhere in the country anytime during the year. Now not only could animals be slaughtered any time of the year, but also the meat could be stored for a long period of time. This meant that meat could be distributed to everyone in the country. This had the added effect of lowering the price of meat. Americans began to enjoy a healthy diet at a relatively low cost.

 ## Crop Production

Advances also were made in the producing of crops. As people moved westward, vast areas of fertile land were opened for farming. A particularly fertile area was the grassy plains of the Midwest. When settlers first came into the area, they had difficulty plowing the soil. The heavy soil could not be turned with the wooden plows used by the settlers. A heavy cast-iron plow would not work because animals could not pull it through the sod (Figure 2–10). The problem was solved in the 1830s by a man named John Deere. He developed a plow with a cutting edge and share made

CAREER DEVELOPMENT EVENTS

The National FFA is an organization for students studying agricultural education in high school. This organization is active in all 50 states and Puerto Rico. Over 450,000 students participate in activities and programs sponsored by the FFA.

This organization began in 1928 when students studying agriculture met in Kansas City for the purpose of establishing a national association. Prior to this, many states had created similar organizations for rural young people. The state organizations began to compete in livestock judging and looked for a place to hold a national competition. The American Royal in Kansas City agreed to host the event. This brought the students together for the first time, and they decided there was a need for a national organization. In 1928 the Future Farmers of America came into existence.

The FFA was patterned after the Future Farmers of Virginia. This state association was developed by Henry Groseclose, Walter Newman, Edmund Magill, and Harry Sanders. They saw the need to give rural youth opportunities to develop social graces and leadership skills. Ceremonies for the FFA were designed after the ceremonies of the Freemasons, and the awards were patterned after the Boy Scouts.

The FFA has passed many milestones since 1928. In 1965 the FFA merged with the New Farmers of America (NFA). The NFA was an organization for black youths studying agriculture. In 1969 girls were allowed to join the organization. In 1988 the name was changed from the Future Farmers of America to the National FFA Organization.

From 1928 until 1998, the FFA held its annual convention in Kansas City. Beginning in 1999, the convention was moved to Louisville, Kentucky. Around 50,000 young

THE HISTORY OF AGRICULTURE

from steel like that used for saw blades. The strong tempered steel allowed the plow to be thinner and lighter (Figure 2–11). Wheels were later mounted to make the plow easier to pull by animals. With this invention, the fertile ground could be plowed under and cultivated.

Another problem was that of harvesting crops. For example, wheat was difficult to harvest because

people spend several days in competition, leadership seminars, award programs, and instructional and motivational programs. This is the world's largest youth convention. The FFA is constantly updating and expanding programs to better serve students. Whether you wish to enter individual competition, team competition, work toward the several degrees of membership, or travel to countries all across the world, there is a program for your needs. There is a place in FFA for you!

The FFA began in 1928 and was organized for students studying agriculture. *Courtesy of the Agricultural Education Program, The University of Georgia.*

FIGURE 2-10
Settlers had difficulty plowing the thick sod of the Midwest.

FIGURE 2-11
The tempered steel of Deere's plow allowed it to be lighter.

FIGURE 2-12
McCormick's reaper made harvesting grain more efficient.

the plants had to be cut by hand. They were tied together into bundles called shocks and stacked. They were then gathered and threshed until the grain left the plants. This took a lot of time and reduced the amount of wheat that could be grown.

In 1831, Cyrus McCormick invented a machine that was used for reaping wheat (Figure 2-12). The horse-drawn implement cut wheat that previously had to be cut by hand. It operated by means of a sickle cutter that was powered by the turning of the wheels. Soon threshing machines were developed to separate the grain from the rest of the plant. This machine was run by steam power, and people fed wheat shocks into the machine and it threshed the wheat grain out. After the invention of the internal combustion engine, the process of cutting the wheat and threshing it was generally combined into one

FIGURE 2-13
A modern combine cuts and threshes the grain.

operation with a machine that came to be known as a combine (Figure 2–13).

The development of the internal combustion engine had a revolutionary effect on agriculture. After its development, work that once took days using human or animal power could now be done in a matter of minutes and with a much better result. Today almost all operations involved in the production of agricultural products are mechanized.

Scientific Research

All of the developments mentioned so far had a real impact on American agriculture. However, by far the most important aspect of the development was scientific research. Scientific research is the systematic search for new knowledge. As mentioned earlier, better ways of growing plants and animals were developed by trial and error. However, this method was too costly and took too long. Dramatic progress was not made until the development of methods of scientific research.

Systematic scientific research began in this country about the middle of the 1800s. It had become apparent that agriculture was a major economic basis for the nation. Leaders thought that our abilities should be strengthened, and the best way to do that was to educate people about agriculture. At that time, the universities in the United States taught a curriculum known as the classics. Students studied subjects such as Latin, Greek, history, philosophy, and mathematics. Though these were challenging disciplines, they had little practical use. People began to realize there was a need for institutions of higher learning where students could study subjects such as agriculture that had a practical application.

In the late 1850s a senator from Vermont, Justin Morrill, introduced a bill to provide public land and funds to establish universities to teach practical methods of producing food and fiber. The bill passed in 1862 and became known as the Land Grant Act, or the Morrill Act. This act provided each state with public land to build a "college for the common people" (Figure 2–14). The purpose of these colleges was to teach agriculture and mechanics. Many of these institutions were called "A&M" universities. Many, such as Texas A&M, still use this name. During that same year, President Lincoln signed into law a bill that established the United States Department of Agriculture (USDA). Soon almost all of the states in the country established Land Grant colleges.

As students enrolled and classes began, a severe problem was recognized. People began to realize they had little knowledge about agriculture. Most of the

FIGURE 2–14

In 1862 Land Grant universities were established.

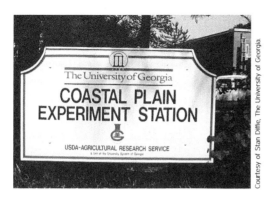

FIGURE 2-15
Experiment stations provided research on new and better ways of production.

knowledge about growing plants and animals had been obtained through trial and error. Most of the ideas about agricultural production had no real scientific basis. To solve this problem, Congress passed the Hatch Act in 1872, which authorized the establishment of experiment stations in states with Land Grant colleges. The purpose was to create new knowledge through a systematic process of scientific investigation (Figure 2–15). These experiment stations put to use what has come to be known as the scientific method of investigation. Here a systematic, scientific investigation about how plants and animals grow was conducted. Most of the progress of American agriculture came about as a result of the research at Land Grant universities and their experiment stations.

In 1890, another Land Grant Act was passed that provided additional colleges. These served black people who lived in the South. These colleges and universities have made a tremendous contribution to the research and teaching of agriculture. Today, the "1890" Land Grant colleges are open to all people.

In 1914, Congress passed the Smith-Lever Act, which set up the Cooperative Extension Service. This legislation placed an extension agent in each county in each state. The purpose was to teach the new information learned from research to people. This completed a system known as the Land Grant concept. This concept was that the purpose of a Land Grant university was to teach, conduct research, and carry the new information to the people in the state through the Extension Service.

In 1917, the Smith-Hughes Act established vocational agriculture in the public high schools as a means of teaching new methods of agriculture. This act established the class you are now in.

Changes in Agriculture

In the past, agriculture has seen many changes. Some of these changes have taken place slowly, whereas others have happened relatively quickly. Modern agriculture is changing at the most rapid pace in history. In fact, you will probably see more change in the way we grow food and fiber than any generation in history.

Scientists are on the verge of being able to locate each of many specific genes on plant and animal chromosomes. In fact, the entire gene sequence of several plants and animals have already been completely mapped. Hundreds of different genetically engineered crops are currently being grown. When scientists are able to map all of the genes on chromosomes, the potential is almost unlimited. Literally thousands of different combinations of gene splices can be accomplished that will make plants and animals easier to grow. In addition, the nutritional value of our food supply will be enhanced. Chapter 24 will explain more about these advancements.

Food products will also change. The trend over the past few years has been toward the sale of processed or semiprocessed food. Consumers want food that is quick and easy to prepare. In the future we will see less and less fresh meat and vegetables sold in the grocery store. Food that can be placed in the microwave for a few minutes and come out as a ready-to-eat dinner will be the way most food is prepared.

There will likely be more changes. Most of the changes we will see during the next 20 years, we can't even imagine. Scientists will find newer and better ways to grow, process, and distribute food and fiber.

SUMMARY

The development of new methods through research and the teaching of this knowledge helped bring about progress in agriculture. Developments such as pesticides, new and better varieties of plants, fertilizers, and soil conservation all resulted from scientific research. As you study the remaining chapters, keep in mind that the agricultural industry is the backbone of our nation. This powerful industry was built as a result of scientific research and development.

CHAPTER REVIEW

Student Learning Activities

1. Interview several elderly people who have lived in your community for many years. Ask them about the importance of agriculture to the development of the community. Report your findings to the class.

2. Pick a particular segment of agriculture and do further research on how that segment developed. Discuss your findings with the class.

3. Contact the cooperative extension agent in your area. Discuss some of the latest agricultural research being done or recently completed at the Land Grant university in your state. Discuss the impact this research might have on agriculture in your area. Report to the class.

4. Research an ancient civilization and report on the role agriculture had on the development of the society. What crops and animals did they grow? What techniques did they use to produce them?

True/False

1. _____ Before any civilization could exist and flourish, a strong agricultural base had to be established.

2. _____ At the time of the American Revolution, 50 percent of the colonists made their living through agriculture.

3. _____ The main crops during the American Revolution were tobacco, sugarcane, and rice.

4. _____ The rich soil and climate made the New World an ideal place for agriculture.

5. _____ Sea Island cotton fibers were extremely difficult to separate from the seeds.

6. _____ Eli Whitney developed a machine to remove seeds from cotton.

7. _____ The northern states developed their economy around industry.

8. _____ President Lincoln signed into law a bill that established the United States Department of Agriculture.

9. _____ Most of the immigrants who first came to the New World were industrialists.

10. _____ Most of the cotton produced went to Europe.

11. _____ Food will probably change very little in the coming years.

12. _____ Genes can be located on the chromosomes by mapping.

Multiple Choice

1. The civilizations that provided a foundation for future civilizations were the Greeks and the
   a. Romans
   b. Americans
   c. Europeans

2. Crops were sent to England and exchanged for
   a. wheat reapers
   b. cotton gins
   c. manufactured goods

3. In colonial times, the most widely grown cotton was
   a. King Cotton
   b. manufactured cotton
   c. Sea Island cotton

4. The invention that allowed the transportation of meat anywhere in the country, anytime during the year, was the refrigerated
   a. cotton gin
   b. boxcar
   c. railhead

5. A machine for reaping wheat was developed by
   a. Cyrus McCormick
   b. Eli Whitney
   c. John Deere

6. The most important aspect in the development of agriculture was
   a. wheat reaping
   b. scientific research
   c. colony industrialization

7. Another name for the Land Grant Act was the
   a. Morrill Act
   b. Hatch Act
   c. Hughes Act

8. The Smith-Hughes Act established
   a. manufactured goods
   b. scientific research
   c. vocational agriculture

9. The biggest problem cattle producers faced was
   a. research
   b. transportation
   c. climate

10. The Smith-Lever Act set up the
    a. Cooperative Extension Service
    b. refrigerated boxcar
    c. Smith-Hughes Act

11. Scientific agricultural research began in this country during
    a. the middle 1700s
    b. the middle 1800s
    c. the middle 1900s

12. The cotton gin
    a. picked cotton
    b. made cloth from cotton fibers
    c. separated fibers and seed

Discussion

1. What problem did cattle producers have?
2. Why is John Deere important?
3. What is the Land Grant Act?
4. What is the Land Grant concept?
5. Why was cotton so important in the South?
6. Why is agriculture important?
7. Why was agriculture necessary to the development of civilizations?
8. How does the cotton gin operate?
9. How did scientific research help develop agriculture?
10. Why did cotton become so important?

CHAPTER 3
Soil: The Origin of Life

Student Objectives

When you have finished studying this chapter, you should be able to:

- Explain why agriculture and all of life depend on the soil.
- Explain the difference between organic and inorganic soil.
- Discuss the carbon cycle.
- Discuss the different ways soils are formed.
- Contrast the differences between acidity and alkalinity and the effect of each on the soil.
- Explain the concept of a soil ecosystem.
- Conceptualize the balance of organisms in the soil.

Key Terms

soil
food chain
humus
organic matter
minerals
glaciers
erosion
floodplain
texture
soil horizon
ecosystem

The earth has a thin coating of material that is called soil. Imagine that the earth is an apple. The layer of soil would be as thin as the peel of the apple. Yet without this layer, there could be little life on the earth. Most plant life is dependent on the soil. Plants are rooted in the soil and get their life-sustaining nutrients from it. If there were no soil, there would be no plants for land animals to eat. Life is in a constant cycle with the earth. A process called the food chain begins with the soil. A plant obtains nutrients from the soil and uses them through energy from the sun. To obtain the nutrients it needs, a mouse may eat

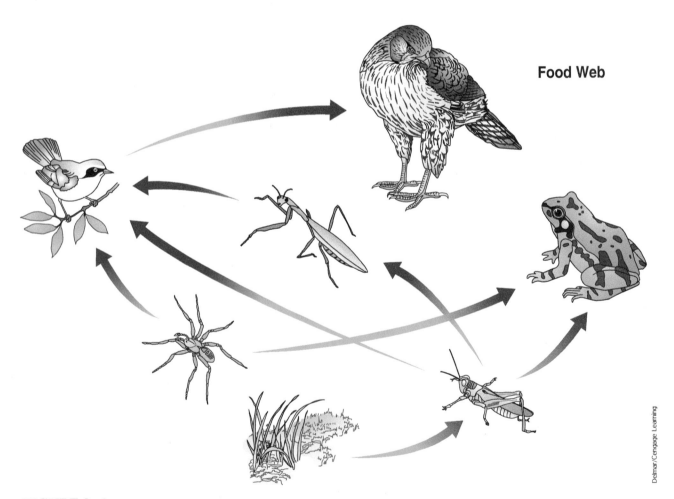

FIGURE 3-1
In the food chain, nutrients are passed from the soil to living creatures and back to the soil.

the plants. An eagle then eats the mouse and obtains nourishment. Eventually the eagle dies and its body is decomposed. The decaying eagle returns nutrients to the soil and the chain begins again. In this process all life is dependent on the soil (Figure 3–1).

Soil is composed of four ingredients: minerals, air, water, and humus. Humus is another word for organic matter, and minerals are substances that did not come from living organisms. Minerals have the chemical and physical properties that can be formed into rocks. Minerals in the soil are very small particles. As these fine particles are grouped together, they form pores or gaps where the particles touch. These pores are filled with either air or water.

Organic Soils

Organic matter comes from living sources. In the example of the eagle that dies, the material left when it decays is organic material, or humus. Soil can be grouped into two very broad categories based on how much minerals or humus the soil contains. Soil that has a lot of humus is called organic soil. This soil is the result of a buildup of plant materials that grew in the soil, died, and decayed. Over many hundreds of years, the decayed plants turned into a black fertile soil called organic soil (Figure 3–2). Organic soils exist in parts of the Midwest where centuries of grasses grew and returned to the soil when they decomposed. Another area with these soils is southern Florida, where swampland has been drained. For thousands of years, a type of grass called saw grass grew in the water. When it matured, it fell into the water and decayed. Over time, several feet of this organic material turned to soil.

Organic soil is very rich in nutrients. Also, the organic material in the soil allows nutrients in fertilizer to be available for plant use. Organic soils are some of the world's most productive soils.

FIGURE 3-2
Organic soil was formed by hundreds of years of decaying plant materials.

Inorganic Soils

Soils that originated from minerals are the result of the wearing down or breaking down of rock. This process is often the result of weather. As wind blows against a rock, particles are loosened and removed (Figure 3–3). Likewise rain, sleet, and snow on rocks can remove particles that become soil. The process of wetting and drying, as well as the dissolving of minerals by the water, can also help in the process of breaking down the stone. The freezing and thawing of water in cracks and crevices of rocks can wear away tiny particles and can open other cracks for the entrance of more water. The process of weathering of rocks is an extremely slow process that occurs over thousands and even millions of years.

Ice may form soil by moving across the surface of the soil. Thousands of years ago, great rivers of slowly moving ice called glaciers extended across the northern portions of America, Europe, and Asia. As these huge masses of ice crept southward, they carried along rocks that were in the path. As

FIGURE 3-3
Wind is a powerful force in weathering rocks and the formation of soil.

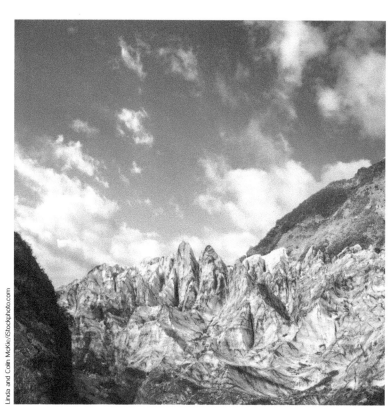

FIGURE 3-4
Slow-moving glaciers crush and pulverize rocks into soil particles.

the glaciers moved along, the rocks were crushed together with enough force that some were pulverized and made into particles small enough to become soils (Figure 3-4). With the change of the seasons, part of the ice would melt and then freeze again. During warm periods, the front of the ice mass would melt and drop all the debris it had picked up along the way. When the weather once again turned cold, the ice resumed its movement and picked up other rocks and soil particles. This mass of large boulders, smaller rocks, sand, silt, and clay is called till. The land mass built up by the dropping of till by a glacier is called a moraine. Obviously, the type of debris picked up by the glacier determined the type of soil that was deposited. Moraines vary in the type and texture of the soil. Some contain a lot of sand and are coarser, whereas some contain the finer silt.

Water-Deposited Soil

As mentioned earlier, moving water wears away rock and forms soil particles. Soil that has been transported and deposited by moving water is called alluvial soil. Some of the world's most productive soils are the result of alluvial action. Water is a major way that soil is moved from one place and deposited in another. Water in a stream picks up soil from the ground. When rains wash soil particles into the stream, the currents carry the material down the stream (Figure 3–5). This process is called erosion. As long as the water is moving rapidly, the soil particles remain suspended in the water. As soon as a streambed begins to level out and the water slows down, the soil particles settle to the bottom. If the stream flows out of its bank, deposits are left in the area covered by the water. These areas are called floodplains. Also, as the stream nears its mouth,

FIGURE 3–5

Water that moves in a stream often picks up and transports soil.

FIGURE 3-6
The rich farmland of the Mississippi Delta is the result of soil transported by water.

the water slows down and the deposits settle to the bottom. These areas are called deltas. Gradually over the years, the soil is built up to a great depth by both of these processes.

In this country, the Mississippi River has carried deposits south for thousands of years and deposited them at the mouth of the river. The result has been the rich farmland of the Mississippi Delta (Figure 3-6). Since the early history of our country, this area has produced rich crops. It should be noted, however, that the soils that made up the Delta were once in the Midwest. The soils in the areas where the soil left are less productive because of the erosion.

Soil may also be deposited into a lake. As a stream empties into a lake, the soil deposits are dropped and they sink to the bottom. Eventually over thousands of years, the lake becomes filled with soil deposits. When the lake disappears, a deep fertile soil is left in place of the water.

Soil Deposited by Wind

The wind also creates and transports soil. In many areas of the world, tall outcroppings of rock have been carved by the wind. As the wind wears away the stone, the loose particles become soil and form deposits known as eolian soils. If the particles are large particles, the soil is called sand and the deposits are called dunes. In parts of the western United States, vast areas are covered in huge sand dunes that are constantly moving and changing shape. Because the sand is constantly moving, little vegetation can grow on the dunes (Figure 3-7).

Fine soil particles such as silt and clay that have been deposited by the wind create soil known as loess soils. These soils may be found through the Mississippi Valley, parts of the Midwest, Washington, and Idaho. Many of these soils were formed when the glaciers melted and the climatic conditions turned dry. When the soil dried out the fine particles were blown by the wind and accumulated in new areas.

FIGURE 3–7
Sand dunes constantly move and little vegetation can grow on them.

In many of the mountainous regions of the world, soil was (and still is) formed by the action of volcanoes. Volcanoes bring deposits from deep within the earth and spew the material onto the surface. Much of this material is in the form of molten rock (called lava). This inorganic material flows out over the surface and cools into solid rock. Other deposits are spewed high into the air. The finer particles are called volcanic ash and may be carried many miles by the wind. Near the volcano many feet of volcanic ash may be built up in a matter of days. Areas as far away as a hundred miles may receive measurable deposits of ash. Soils from volcanic ash are generally very productive soils (Figure 3–8). Volcanic soils can be found in the states of Hawaii, Oregon, and Washington.

FIGURE 3–8
Soil from volcanic ash is usually very fertile.

SOIL: THE ORIGIN OF LIFE

Soil Texture

The **texture** of the soil refers to the size of the individual soil particles. The larger the size of the soil particles, the coarser the soil feels when rubbed between the fingers. The largest of the soil particles is sand. Sand particles are small but they are considered large compared to other types of particles. If a soil containing sand is rubbed between your fingers, it will feel coarse and gritty, like corn meal. Silt particles are smaller than sand and feel smoother to the touch. The smallest of the particles is referred to as clay. A soil high in clay content will feel smooth, like flour. One way to illustrate the relative sizes is to imagine that if a clay particle were the size of a BB, a silt particle would be the size of a golf ball, and a sand particle would be the size of a basketball (Figure 3–9).

Soils are seldom composed of pure sand, pure silt, or pure clay. More often, they are a combination of the three. Soil that contains less than 52 percent sand, 28 to 50 percent silt, and 7 to 27 percent clay

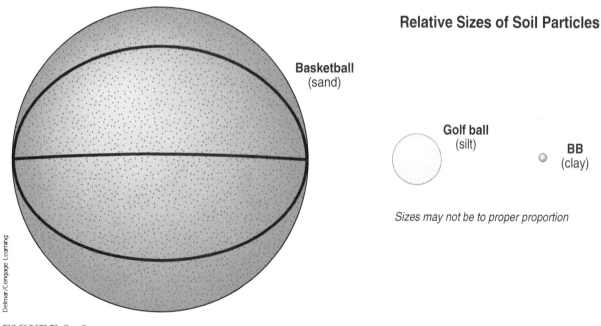

FIGURE 3–9
If a clay particle were the size of a B, a silt particle would be the size of a golf ball, and a sand particle would be the size of a basketball.

CAREER DEVELOPMENT EVENTS

The FFA Land Judging Career Development Event is a competition for teams of four members. Each member competes individually and the final team score consists of the total of the three highest individual scores. The FFA land judging contest aids classroom instruction by teaching the students practical application of the Soil Conservation Service classification system. The competition and hands-on applications motivate students to learn soil and water management practices. A student's decision-making skills are developed and environmental awareness increased through the analysis of soil characteristics and the identification of important soil and water resources.

This event tests the FFA members' ability to determine a site's topsoil texture, thickness, and effective depth of topsoil and subsoil. The student must also decide the permeability of subsoil, land slope, erosion, drainage, and any other factors that will best decide the land's capability. After the student determines land class, the proper land use and treatment practices must be decided. The student judges the land by the observation and measurement of prepared soil pits and the land area surrounding each pit. Each team member should come to the event equipped with a soil probe, pencil, clipboard, and calculator. After all team members have completed each phase of the event, each individual score will be calculated. Winning teams from each state advance to the National Land Judging Career Development Event.

These activities help make classes more interesting and give students an additional reason for doing their best to develop necessary skills. The Land Judging Career Development Event requires students to study classroom material and practice the hands-on activities. The extra time in preparation for the event not only helps produce a winning team but also helps prepare for a career.

The Soil and Water Management Proficiency offers individual FFA members an opportunity to win an award for experiences related to renewable natural resources. Students who have participated in the planning and use management practices to improve air, soil, water, timber, fish, or wildlife can apply for this award. Programs in which a member is involved in helping to stimulate public awareness or assisting in educating the public concerning pollution problems should be included in the application. Other activities a student might participate in are the development and implementation of management practices that will prevent erosion, improve the productivity of the soil, promote efficient use of water, or reduce air and water pollution. To compete for the proficiency, the student should keep a list of related skills developed as a result of the above activities and training in vocational agriculture. The student should participate in a Supervised Agricultural Experience (SAE) that relates to soil and water management. Proficiency applications are judged on the student's records, activities, and experiences. Keeping accurate records of accomplishments and soil and water management-related income and expenses are essential to becoming a winner.

In the Land Judging Career Development Event, students judge the land by the observation and measurement of prepared soil pits and the land surrounding the pits.
Courtesy of Frank Flanders

FIGURE 3-10
Soil with a high clay content does not allow water to pass through very fast. These soils can be very productive.

is called loam soil. This soil is considered to be near ideal for growing most crops. Soils that have a large concentration of sand are coarse and do not hold water very well. Because the particles are so large, the water from a rain or irrigation system passes on through the soil. Water that goes on through the soil is of little use to plants. Clays and silts slow the water down as it is absorbed into the ground and hold a portion of it that the plants can use (Figure 3–10). If the soil has a large percentage of clay, the ground will not let enough water through. This may either cause the water to run off or may hold the water too long and cause problems for the plant.

Clay also plays an important role in soil fertility. Most producers apply fertilizers to the soil. Fertilizers provide the nutrients needed for the plant to grow. Clay particles in the soil hold the nutrients in place until the plants can use them.

Soil pH

The pH of the soil is a measure of how acidic it is (Figure 3–11). Acids contain hydrogen that forms particles called ions when dissolved in water. Alkaline is the opposite of acid. Vinegar is a common substance that is acidic. Baking soda is alkaline. The pH scale goes from 1 to 14. A score of 7 means that the substance is neutral. The lower the number, the more acid the substance is. The higher the number, the more alkaline the substance.

Certain crops may have problems growing in soils that are too acidic or too alkaline. Producers frequently adjust their soils to the right pH. They have samples of their soil tested for pH. If the soil is too acidic, the correct amount of an alkaline substance (such as lime) is added. If the soil is too alkaline, acid materials (usually sulfur-based materials) are added to the soil. Crops such as blueberries grow well in soils that are relatively acidic (pH of 4.0 to 5.0). Crops such as asparagus do well in soils that are less acidic (pH of 6.0 to 8.0).

SOIL: THE ORIGIN OF LIFE

FIGURE 3-11
The pH scale measures the acidity or alkalinity of a substance.

 Soil Horizons

When soil is formed from rock or from organic material, the soil forms in layers. These layers lie parallel to the surface of the earth. As layers are added, they begin to take on different characteristics. These different regions are called soil horizons (Figure 3-12). If a pit is dug deep into the ground, the vertical bank on the side of the hole gives a picture of the different layers, or horizons, of soil. This picture is called the soil profile (Figure 3-13).

The uppermost layer is called the O horizon. This layer contains undecomposed and decomposed organic matter. This is the place where most seeds germinate.

The next layer is the A horizon and is composed of the topsoil that contains organic matter along with minerals. It is generally darker than the horizons below and may be from a few inches to several feet thick.

The region under the A horizon is the E horizon and is the area of the topsoil that has little organic material. This horizon is light in color because the clays and humus have been washed out by water that passes through the soil.

The next region is the B horizon. In this subsoil region, materials washed out (or leached) from the other horizons accumulate and form. This area is generally high in clay content. The B horizon combined with the O, A, and E horizons make up a region of the soil where plant roots grow.

Underneath the B horizon is the C horizon that is made up of parent material. Parent material is the material from which the soil originated. This horizon has not been affected by the soil-making process and does not have the properties of the soil in the other horizons.

The last horizon in the soil profile is the R horizon. This horizon is composed of the bedrock on which the other soil horizons rest.

FIGURE 3-12
Soil is classified in soil series. Some examples are: (A) Entisols. (B) Aridisols. (C) Iceptisols. (D) Aflisols.

SOIL: THE ORIGIN OF LIFE

FIGURE 3-13
A pit dug into the ground can give a view of the soil profile.

 The Soil Ecosystem

Your first observation of the soil might lead you to believe that it is lifeless. Looking across a freshly plowed field, you might think that all that is there are minerals and decaying plant matter. Actually, the soil abounds with life. The soil is an ecosystem. An ecosystem is all of the plant and animal life that live in an area. The life-forms depend on each other for the proper balance of food and other environmental factors. Within this ecosystem are many different forms of plant, animal, and microbial life. Most of these forms of life depend on each other for their existence or for a maintenance of the proper balance in nature.

Plant Life

Most plants depend on the soil for their existence. The soil provides the support for the root system that anchors the plant and makes it stable. Most of the nutrients used by the plant are obtained from the soil. Even though we are used to seeing the top portions of plants, a very large portion of the plant lives in the soil (Figure 3-14). Roots of most plants can reach as far as several feet into the soil. In fact,

FIGURE 3-14
A large part of a plant lives in the soil.

such common crops as alfalfa have been known to grow as far as 25 feet or more into the ground. The area of the soil that contains the roots of plants is called the rhizosphere. In this zone of the soil the plant receives water and the nutrients it needs to live and grow.

Microorganisms

Within the rhizosphere live billions of microorganisms of many different types. Many of these organisms live off the roots of plants. As plants live and grow, they are constantly oozing materials out through the roots. This material contains protein and other materials that the microorganisms feed on. In addition, as root cells mature and die, the microorganisms act to decompose the cells. When the entire plant dies they help to return the nutrient material the plant took from the soil back to the soil. This is known as the carbon cycle.

The most common soil microorganisms are bacteria. They are so abundant that in only one teaspoon of soil there can be as many as 500 million bacteria. Several types of these bacteria live in what is known as a symbiotic relationship with plants. This means that organisms of different types live together for mutual benefit. The most common symbiotic bacteria are the nitrogen-fixing bacteria, rhizobia, that live in the roots of certain plants. Plants that host these organisms are called legumes and include such plants as beans, clovers, peanuts, alfalfa, and peas. Rhizobia live in lumps on the roots called nodules. The bacteria live in the soil and attach themselves to the root hairs of the legume shortly after the plant sprouts. As a reaction to the bacteria attaching itself to the root, a gall, or knot, forms on the plant root. It is within this swelling that the rhizobia bacteria grow and reproduce. They receive all the nutrients they need to live and reproduce from the host plant. In return the bacteria convert nitrogen from the air in the soil into a form of nitrogen that the plant can use (Figure 3–15).

FIGURE 3-15
Bacteria, called rhizobia, form nodules on roots. They convert nitrogen from the air to a form useful to the plant.

FIGURE 3-16
Soil fungi help to decompose plant material.

Fungi are plantlike organisms that contain no chlorophyll. They range in size from microscopic to the large mushroom fungi that grow on the surface of the soil or on decaying plant material. These organisms are as abundant as bacteria in the soil. They play an important role in the breakdown and decay of plant materials (Figure 3-16). Fungi are particularly important in forest soils because they break down lignin, which is a primary component of wood. Trees that die and fall to the ground are returned to the soil through the action of fungi, bacteria, and other organisms.

Protozoa are one-celled organisms that live in moist soil. They are aquatic organisms because they live in particles of water in the soil. When the soil becomes dry, the protozoa change into an inactive state until the soil becomes moist again. Protozoa feed on bacteria in the soil. Through this process, a balance of bacterial life in the soil is better maintained.

One of the most important groups of microscopic animals that inhabit the soil is nematodes. Nematodes are worms in the class Nematoda that have smooth round bodies that are not segmented.

Although many nematodes are not microscopic, most of those that live in the soil are so small they can't be seen with the naked eye. These tiny worms are very abundant. In a typical spade full of moist soil there may be over a million nematodes. In fact, they are the most abundant multicelled animals in the soil (Figure 3–17). There are three basic groups of soil nematodes. One group consumes decaying organic matter. Another group eats other microorganisms. The third group feeds on plant roots and is by far the most important because of the damage they do to the plants.

FIGURE 3–17
Nematodes can feed on the roots of plants. They are the most numerous multicelled animals in the soil.

SUMMARY

Soil is essential to all life. All of our food sources depend on it. Humans must protect it because it takes hundreds or even thousands of years to build soil. The fertile soil of this nation has provided much of our wealth. If we take care of our soil, our wealth will continue.

CHAPTER REVIEW

※ Student Learning Activities

1. Bring in a sample of soil from around your home. Determine if the soil is organic or inorganic. Also determine the soil texture and if the soil is mostly sand, clay, silt, or loam.
2. Observe the area around your school. Attempt to determine how the soil might have been formed in the area. What indications did you use?
3. Make a list of all the ways soil is important to you. Compare your list to others in the class.
4. Locate soils that are different colors. Determine what makes the difference in color.

SOIL: THE ORIGIN OF LIFE

❋ True/False

1. _____ Most plant life is dependent on the soil.
2. _____ If there were no soil, there would be no plants for land animals to eat.
3. _____ A process called the food chain begins with the soil.
4. _____ Minerals have the chemicals and physical properties that can be formed into soil.
5. _____ Organic soil is rich in nutrients.
6. _____ The largest of the soil products is sand.
7. _____ If the soil is too alkaline, acid materials are added to the soil.
8. _____ The most common soil microorganisms are the bacteria.
9. _____ Soil is made up of only minerals and decaying plant matter.
10. _____ Roots of most plants cannot reach very far into the soil.

❋ Multiple Choice

1. A plant obtains nutrients from the soil and uses them through energy from
    a. water
    b. food
    c. sunshine

2. Humus is another word for
    a. organic matter
    b. bacteria
    c. decaying plants

3. Soil that has been transported and deposited by moving water is called
    a. fertile soil
    b. alluvial soil
    c. eolian soil

4. Fine soil particles like silt and clay that have been deposited by the wind create soil known as
    a. eolian soil
    b. alluvial soil
    c. loess soil

5. In many of the mountainous regions of the world, soil was formed by the action of
   a. volcanoes
   b. earthquakes
   c. neither a nor b

6. Fertilizers provide plants with
   a. insect repellent
   b. nutrients
   c. food

7. Plants that host rhizobia are called
   a. legumes
   b. fungi
   c. nematodes

8. One-celled organisms that live in moist soil are called
   a. protozoa
   b. nematodes
   c. fungi

9. Plantlike organisms that contain no chlorophyll are called
   a. protozoa
   b. legumes
   c. fungi

10. The area of the soil that contains the roots of plants is called the
    a. subsoil
    b. rhizosphere
    c. C horizon

Discussion

1. Soil is composed of what materials?
2. Why is plant life dependent on soil?
3. How is soil classified?
4. List two ways soil can be deposited.
5. What is loam soil?
6. What is the problem with too much clay in soil?
7. What are soil horizons?
8. What is the carbon cycle?
9. What is a symbiotic relationship?
10. What are the three basic groups of nematodes?

CHAPTER 4
Plant Structures and Their Uses

Student Objectives

When you have finished studying this chapter, you should be able to:
- Describe the functions of plant stems.
- Tell the functions of xylem and phloem.
- Explain how stems are used by people.
- Describe the functions of a leaf.
- Explain how leaves are used by humans.
- Discuss the function of flowers.
- Explain how flowers are used by people.
- Explain how bees are used in agriculture.
- Describe the functions of seeds.
- Explain how seeds are used by people.
- Discuss the functions of roots.
- Tell how roots are used by people.

Key Terms

stems
photosynthesis
flowers
nectar
seeds
roots

FIGURE 4-1
Plants provide food for humans and feed for animals raised for food.

All life as we know it on earth is dependent on plants. Without plants, the atmosphere of our planet would not have the oxygen required to support animal life. As well as providing air to breathe, plants are essential for food, fiber, erosion control, and many other necessities. Almost all of agriculture is dependent on plants. Plants provide food for humans and the agricultural animals raised for food (Figure 4–1). Plants, like animals, have different parts that serve different functions. People make use of these plant parts in a variety of ways.

Plants are made up of stems, leaves, flowers, seeds, and roots. Each of these parts plays an important role in the life of the plant. Agricultural producers grow particular plants in order to obtain plant parts or structures that can be used.

Stems

Stems are very important to plants. They support the plant and provide a means of moving nutrients from one end of the plant to the other. Some stems grow into large trunks, as in trees. Other stems may grow in the form of winding vines that allow the plant to climb fences or other plants.

At the very tip of the stem is a small knob, called the terminal tip, containing the bud that will become next season's growth. Farther down the stem there are small scars that go all the way around the stem. These mark the place on the stem where last year's terminal tip, or bud, was. These scars are called bud scale scars. It is easy to calculate how old a stem is by counting these scars (Figure 4–2).

Every place along the stem where leaves are or were has scars, too. These are called leaf scars and they occur at nodes. Also at the nodes are small buds called axillary buds. These buds may at some time grow into stems or side branches. They grow in the axil formed by the leaf and stem (Figure 4–3).

PLANT STRUCTURES AND THEIR USES

FIGURE 4-2
The age of a stem may be calculated by counting the bud scale scars.

FIGURE 4-3
Parts of a stem.

Inside the stem are tubelike structures called the xylem and the phloem. The xylem carries the water and nutrients absorbed by the roots to the leaves. The phloem carries sugars manufactured in the leaves down to the roots. Agriculture makes use of these parts. In trees, the woody part is the xylem. Lumber cut from trees is composed of the xylem of the stems. Power poles, fence poles, and plywood are composed of stem materials. Also, the xylem of stems from trees provides fiber for making paper (Figure 4–4). Just think of all the uses for paper and paper products.

In a plant called flax, fibers are removed and woven into a cloth called linen. These fibers are the phloem of the stems of the flax plant. The outer layers of woody tissue (the xylem) are removed and the phloem is harvested.

Stems are also used extensively for food. The celery we eat in salads or as an ingredient in cooked foods is a stem. Rhubarb and asparagus are also examples of stems that are used for human food

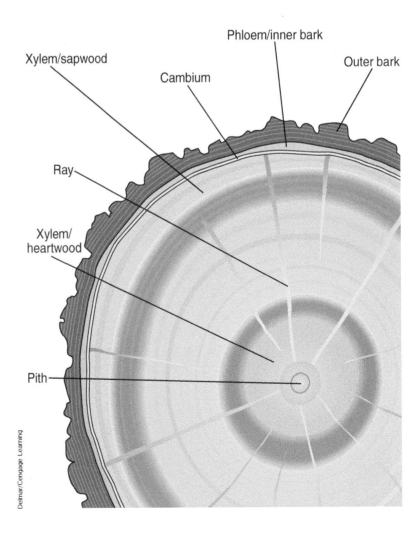

FIGURE 4–4
The xylem of trees is used to make paper.

PLANT STRUCTURES AND THEIR USES

FIGURE 4-5
Stems of plants like asparagus are used for food.

(Figure 4-5). Potatoes are also a type of stem. These stems, called modified stems, grow underground. They provide a place for the plant to store nutrients. These nutrients, stored in the form of starches, have fed people for thousands of years.

Stems can also be used as animal feed. Plants such as alfalfa are ground up and fed to cattle. Although not as nutritious as the leaves, stems provide a food source. Corn stalks and grain sorghum stalks (along with the rest of the plant) are chopped and made into a feed called silage. Many cattle are fed on silage (Figure 4-6).

Leaves

Everyone is familiar with leaves. They provide the beautiful green colors of spring and the reds, yellows, and browns of autumn. Leaves add color and decoration both inside and outside our homes. Potted plants with unusual leaf colors are used inside the home for decoration. Plants that turn different shades and different colors with the seasons are used outside.

Leaves also come in a wide assortment of shapes and sizes. Size, shape, and patterns of leaves are used to identify plants. The leaves of a juniper are less than 1/4 inch long, but the leaves of elephant ear plants can be as large as 3 feet across (Figure 4-7). Leaves

FIGURE 4-6
Cattle eat plant stems in the form of silage.

FIGURE 4-7
Some plants like the elephant ear have huge leaves.

grow in alternate, opposite, or whorled patterns. In the alternate arrangement, leaves appear on one side of the stem at a node and from the other side at the next node. Opposite leaves grow from opposite sides of the stem directly across from each other. Whorled leaves refer to three or more leaves that grow around the stem at the same node (Figure 4-8).

Leaves are where the manufacture of food takes place. In a process known as photosynthesis, the energy of sunlight is used to produce nutrients. These nutrients are used by the plant for growth, maintenance, and reproduction. This food is in the form of sugars and carbohydrates as well as fats and proteins. Because this food is stored in leaves, animals use the leaves for food. Over the years, agriculturists have discovered and developed plants with leaves that are high in nutritive value and that livestock will eat. Cattle, sheep, and horses eat the leaves of grass and other plants as forage. Producers cut and dry forage into hay that is stored and used as livestock feed.

Humans also eat the leaves of plants (Figure 4-9). As with plants used to feed animals, scientists have developed those plants that make good food for humans. Cabbage, mustard, lettuce, and spinach are all examples of plant leaves eaten by humans. The leaves are the primary place for food storage in the plant. The food stored by the plant can be used as food for animals. Many of the nutrients needed by animals and humans are obtained from the eating of plant leaves.

FIGURE 4-8
Leaves of different plants are arranged differently on the stem.

Alternate

Opposite

Whorled

PLANT STRUCTURES AND THEIR USES

FIGURE 4-9
Humans also make use of plant leaves for food.

Flowers

Flowers are the most attractive part of the plant. Most flowers are flashy, like the rose or daisy. Others are so plain and simple, it is hard to realize they are flowers. Flowers serve as the reproductive organs for many plants. Some plants reproduce asexually. This means that these plants reproduce without a sperm and egg uniting. Usually, such plants do not have flowers. Plants that reproduce sexually have flowers.

Some species of plants have male flowers and female flowers on one plant. Others have female flowers on one plant and male flowers on another plant. Most, however, have both the male and the female parts in the same flower.

Most flowers consist of the same basic parts. The female part, called the pistil, is made of the stigma, style, and ovary. The male part is called the stamen and contains the anther and filament. The pistil produces eggs that are fertilized by the pollen produced in the stamen. Pollen is usually taken from the anther to the stigma by insects, the wind, or in some cases, animals (Figure 4-10).

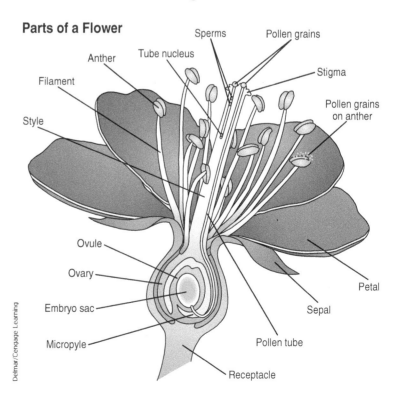

FIGURE 4-10
The parts of a flower.

FIGURE 4-11
Brightly colored flowers attract insects that pollinate the plants.

In order to attract the bees or other insects, plants have developed brilliantly colored flower petals (Figure 4-11). Although insects probably don't see the colors of the flowers the same way as we do, they are attracted by the petals. The flowers secrete a sweet fluid known as **nectar** that the insects use for food. As they crawl or fly into a flower to collect nectar, grains of pollen are caught on their legs or other body parts. When they go to the next flower, the grains of pollen are knocked off and remain in the flower. This causes pollination to occur.

Honeybees are particularly adept at pollinating (Figure 4-12). Many insects work flowers, but most go from one kind of flower to different kinds of flowers each day. Bees, on the other hand, work a particular kind of flower for a period of time. For example, honeybees will work apple blossoms for several days until the flowers are gone, then work a different type of flower. By doing this they go from an apple blossom to another apple blossom and spread pollen from one apple blossom to another. This process ensures that the blossoms are thoroughly pollinated.

Fruit growers hire beekeepers to bring in truckloads of bees in the spring as the trees are blooming. The bees live in wooden, boxlike structures called hives, with a separate colony of bees in each hive. The hives are easy to handle and can be loaded on

FIGURE 4-12
Honeybees are particularly adept at pollinating.

PLANT STRUCTURES AND THEIR USES

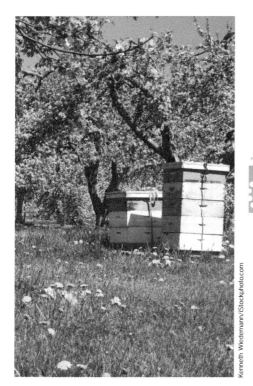

FIGURE 4-13
Commercial honey producers place bees in an orchard to pollinate the trees and to produce honey.

FIGURE 4-14
Breakfast cereals are made from plant seeds.

a truck with the bees still in the hive. The owner of the bees can then move the hives from orchard to orchard for a fee from the fruit or crop producer. In addition, the beekeeper can harvest hundreds of pounds of honey each year that can be sold at a profit (Figure 4-13).

Seeds

Seeds generally result from the flowers on plants. Seeds are another part of the reproductive process of the plant. Most plants begin their growth from seeds. The production of seed is one of the most important links in the entire agricultural industry. In fact, a lot of our food is derived from seed. Estimates are that around 70 percent of our diet comes directly from the seeds of plants. The bread or bowl of cereal that you had for breakfast this morning was made from the seed of grain plants (Figure 4-14). Cooking oil is derived from the oil that is pressed from seeds of plants such as safflower, soybean, or cotton. Snack foods such as peanuts and popcorn are all made from the seeds of plants.

Seeds are also used for a variety of other purposes. Many of our beverages, such as coffee and cocoa, come from the seeds of trees and shrubs. Oils and dryers in paints and wood finishes come from seeds. Oil from the seed of the castor bean was once used for medicine and is now used as an oil in hydraulic systems. Seeds are also used in the manufacture of cosmetics, ointments, and pharmaceuticals.

Seeds produced on plants often are encased in a fruit. For example, apple seeds develop in the pulpy sweet material that is eaten by people and animals (Figure 4-15). Many different fruits encase seed. Plums, peaches, pears, and oranges all have seeds at the center. Some foods we don't generally think of as fruits also fit this description. Tomatoes, squash, eggplant, and watermelons all encase seed and are eaten by people.

FIGURE 4-15
The fruit of an apple encases seeds.

FOOD FOR AMERICA PROGRAM

The Food for America program is an excellent way to expose your FFA to elementary school children. This program is designed to make young students aware of the products and opportunities of agriculture. Many elementary students have never had the opportunity to visit any agriculture industries, businesses, or farms. The agricultural source of many everyday products, from milk to shoe polish, is easily overlooked.

FFA members help elementary students understand where food comes from. *Courtesy of Blane Marable, Morgan Co. High School, Madison, GA.*

PLANT STRUCTURES AND THEIR USES

Roots

The two main functions of roots are to support the plant and to take in water and nutrients from the soil. Plant roots are at least as extensive as its stems and leaves. When you see a large tree just think of how much of it is really in the ground, hidden from your view.

There are four main types of roots: fibrous, tap, adventitious, and aerial. Fibrous root systems have

Few Americans have an accurate understanding of modern agriculture. Many of today's consumers believe that since the number of farmers is decreasing that agriculture is dying. What they fail to realize is that agriculture technology has helped farmers become more productive. Today's agricultural producer uses less land to provide for more people. The Food for America program exposes the network of the many industries associated with agriculture, from production, transportation, and storage to the marketing of food and other agricultural products. Careers as scientists, marketing specialists, engineers, veterinarians, nutritionists, and other occupations are important parts of American agribusiness and should be discussed with elementary students. Through the Food for America program, FFA members help elementary students understand these crucial roles and how food and other agricultural products are produced. FFA members should help elementary school teachers in agricultural instruction and bring examples of animals and agriculture products into the classroom. FFA members may even offer tours of a working farm as a field trip for the elementary students. FFA members are excellent role models. With well-practiced presentations, FFA members can leave a positive image of today's agriculture and the FFA through the Food for America program.

FIGURE 4-16
Humans also make use of roots, such as sweet potatoes, for food.

many tiny roots that lead from the plant deep into the ground. One corn plant can have as much as 100 miles of roots.

Most trees have taproots. These are very large tough roots that reach deep into the earth for moisture and keep the tree from falling over. Carrots are also examples of a taproot system. There are many smaller roots that come off the main root.

Adventitious roots usually occur on plants that also have fibrous or taproot systems. An example is corn plants that have roots coming out of the main stalk above the ground that help to brace the plant against strong winds.

Aerial roots are common in very wet areas of the world. These roots are able to obtain water directly from the air. Plants with this type of root system don't even need soil to grow in. Both orchids and staghorn ferns have aerial roots.

As well as being a necessary part of the plant, roots are important to humans. Plants are often used to slow down or stop erosion. Roots help to hold soil in place. All of us have eaten roots at one time or other. Carrots, turnips, radishes, and sweet potatoes are all examples of roots that are commonly eaten (Figure 4–16). Some spices also come from roots. Ginger is a root that is grown in India and Southeast Asia. It has been used for hundreds of years to add spice to foods.

SUMMARY

Plants have always been used by people to eat, build shelter, and make clothing. Almost all of agriculture is involved with plants. Without plants and their various uses, people would have a difficult time existing. Agriculture has improved plants that were once found in the wild. Chosen for a particular use, these plants were bred to be more efficient. Modern agriculture will continue to produce these plants to fill the needs of people.

CHAPTER 4 REVIEW

PLANT STRUCTURES AND THEIR USES

❋ Student Learning Activities

1. Make a list of all the food in your house that comes from plants. Determine the part of the plant from which the food was made. Share the results with the class.
2. Look around your home and list all the plants that are ornamental. What parts of the plants provide the beauty?
3. Choose a particular plant part (such as leaves, stems, roots, or flowers) and create a list of all the different uses that could be made of this part.
4. Locate several different types of flowers. Dissect them and identify the parts.
5. Bring in at least five products made from seed. Report to the class on how seeds were used in making the products.

❋ True/False

1. _____ Potatoes are a type of stem.
2. _____ Plants reproduce asexually.
3. _____ Leaves serve as the reproductive organs for many plants.
4. _____ Some species of plants have male flowers and female flowers on one plant.
5. _____ Nectar is a pesticide.
6. _____ Aerial roots are common in cold areas.
7. _____ Plants are often used to slow down or stop erosion.
8. _____ Whorled leaves refer to two or more leaves that grow around the stem at the same node.
9. _____ Leaves are where the manufacture of food takes place.
10. _____ In a process known as photosynthesis, the energy of sunlight is used to produce nutrients.

❋ Multiple Choice

1. At the very tip of the stem is a small knob called the
   a. terminal tip
   b. node
   c. xylem

2. The structure inside the stem that carries the water and nutrients from the roots to the leaves is the
    a. phloem
    b. node
    c. xylem

3. Flowers are really just specialized
    a. leaves
    b. plants
    c. vegetation

4. Those plants that reproduce sexually have
    a. leaves
    b. flowers
    c. stems

5. Honeybees are particularly adept at
    a. asexual reproduction
    b. producing nectar
    c. pollinating

6. Most tree roots are
    a. fibrous
    b. tap
    c. aerial

7. The xylem of tree stems provides fiber for making
    a. cotton
    b. linen
    c. paper

8. The male part of the plant is called the
    a. pistil
    b. stamen
    c. stigma

9. Insects are attracted to the
    a. petals
    b. leaves
    c. color

10. Seeds produced on plants often are encased in
    a. plants
    b. soil
    c. fruit

Discussion

1. Why are plants so essential to us?
2. What are the major parts of plants?
3. What are the functions of the xylem and phloem?
4. What are some uses humans make of plant stems?
5. What are the parts of the flower?
6. Why are insects important to flowers?
7. List the main types of roots.
8. Why is the production of seed one of the most important links in the agriculture industry?
9. For what purposes can leaves be used?
10. How are bees used in agriculture?

CHAPTER 5
Agricultural Pests

Student Objectives

When you have finished studying this chapter, you should be able to:

- Explain why producers have to deal with pests.
- Discuss how insects are harmful.
- Define the life cycle of insects.
- Discuss how insects are controlled.
- Analyze the concept of integrated pest management.
- Explain what a weed is.
- Tell how weeds are controlled.
- Discuss the newest methods used in pest control.

Key Terms

insects
pests
disease
parasites
insecticides
integrated pest management
biological control
weeds
herbicide

Agriculture deals with the growing and cultivating of living things for the use of humans. People provide the best care possible for the plants and animals they grow. One of the basic parts of caring for crops and animals is protecting them from other living things. In the wild, plants and animals are constantly in competition with each other. In nature, all living things need space and nourishment. Animals eat other animals, animals eat plants, and plants get nutrients from the soil. When humans grow plants or animals, they provide an enticing meal for animals such as insects. Every year, cropland is prepared for planting. The soil is tilled, fertilizer is spread, and conditions are made favorable for the crop to grow. Unfortunately, the conditions also favor the growth of unwanted plants.

Many industries have developed to protect plants and animals from pests, such as the manufacture of cultivation equipment and the development of agricultural chemicals (Figure 5–1). These industries use a tremendous amount of resources each year. Machines and chemicals have to not only be made but also have to be distributed to the producer.

Insects

Insects are what most people think of when they think of agricultural pests. Since humans first began growing their own food, insects have been a problem.

FIGURE 5–1
A vast industry has developed to protect plants from pests.

AGRICULTURAL PESTS

Many ancient documents tell of troubles with insects, such as locusts, that devoured crops. For thousands of years it was accepted that insects would get a large percentage of the crops (Figure 5–2). Problems with insects continue even to this day. If insects are not controlled on crops, millions of dollars of damage will be done. Not only will insects eat a large part of the produce, but they will also damage most of the crop not eaten. We are accustomed to seeing crisp, fresh apples at the grocery store. These apples are almost flawless and have no damage. However, if insects are not controlled as the apples are growing, the apples we buy will have defects such as wormholes. Most consumers will not buy this type of apple (Figure 5–3). The same is true of almost all that we grow.

FIGURE 5–2
In the past it was accepted that insects would get a large percentage of the crops. Insects such as this potato beetle still cause damage.

FIGURE 5–3
Consumers are reluctant to buy insect- or disease-damaged fruit.

FIGURE 5-4
Insects such as mosquitoes carry disease.

Not only do insects eat our crops, but they also carry disease. Millions of people have contracted diseases because of insects. Insects such as the mosquito may draw blood from a person infected with a disease and then inject disease organisms into the next person it bites. If the same mosquito bites many people, the disease becomes widespread (Figure 5-4). Yellow fever, malaria, West Nile virus, and encephalitis are diseases spread by mosquitoes. Throughout history, countless millions of head of livestock have died as a result of diseases carried by insects.

Some insects are parasites. Parasites are organisms that live on other organisms (called hosts) and harm them. Agricultural animals are harmed by insect parasites such as ticks, fleas, and heel flies. These insects get their nutrients from the animal they live on or in. This causes the animal to have to eat more in order to live and do well. In turn, the producer must give more feed to the animal or control the parasite.

One of the biggest problems with insects is that there are so many of them. Over 750,000 species have been classified. Scientists think there may be as many as 2,000,000 more that have not been identified. Of the identified insects, only about 500 are considered to be major pests in this country. However, these pests are formidable enemies. One female insect may lay millions of eggs during her life (Figure 5-5).

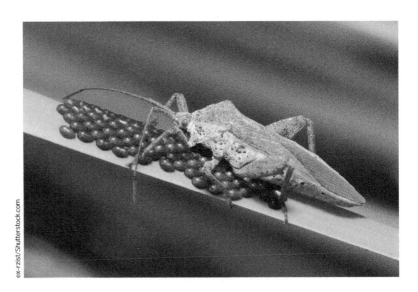

FIGURE 5-5
One female insect may lay millions of eggs in her lifetime. This insect is a stinkbug.

AGRICULTURAL PESTS

If most of these hatch, there will be a lot of insects around to cause trouble. Fortunately, nature helps out. Many birds and animals eat large amounts of insects. Other insects such as the lady beetle and the praying mantis devour huge numbers of insects. Each year, vast numbers of insects are eaten by predators.

Even with all the predators eating insects, that still leaves a lot to destroy crops. For many years we have used chemicals to get rid of insects. These chemicals, called insecticides, have been very efficient. They have been the most efficient means ever devised to kill insects. In the past, insecticides have been used in tremendous quantities. The overuse of insecticides has resulted in problems. Many of the older chemicals such as DDT have been harmful to the environment, and wildlife has suffered. These chemicals remain in the soil for many years. Rain would wash the chemicals into streams where they were absorbed by aquatic plants and animals. Animals such as water birds and eagles ate the fish and absorbed the chemicals. Reproductive cycles were disrupted because the eggs from the birds would not hatch (Figure 5-6).

Today, all insecticides are regulated very closely by governmental agencies. Most of the old insecticides that remained in the soil for so long have been

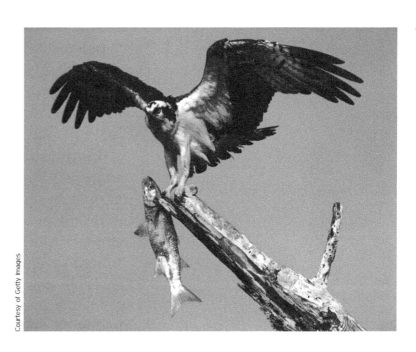

FIGURE 5-6

Reproductive cycles of animals such as water birds and eagles were disrupted by DDT.

FIGURE 5-7

Insects, such as the lady beetle, devour other insects.

banned. New insecticides break down into harmless substances in a short time. This makes environmental problems less likely because the residues of the chemicals do not build up.

Any sound insect control program involves **integrated pest management** (IPM). This approach combines knowledge of the pest and its life cycle, monitoring, cultural practices, biological control, and chemical pesticides.

Much has been done recently in the area of **biological control.** Biological control involves the use of predatory insects, fungi, or viruses. You are probably familiar with the appetite of insects such as the praying mantis and the lady beetle (Figure 5-7). In some areas, these insects are raised for the purpose of eating insects on crops.

Cultural practices have a big impact on diseases and insects. Simply pruning out dead and diseased limbs makes a difference. Burning these diseased limbs and weeding around plants helps a lot. This helps to break up the life cycle of insects. Insects go through several phases in their lives (Figure 5-8). The cycle begins when the female lays eggs. The eggs hatch into wormlike organisms called larvae. Most of the harm insects do comes about during

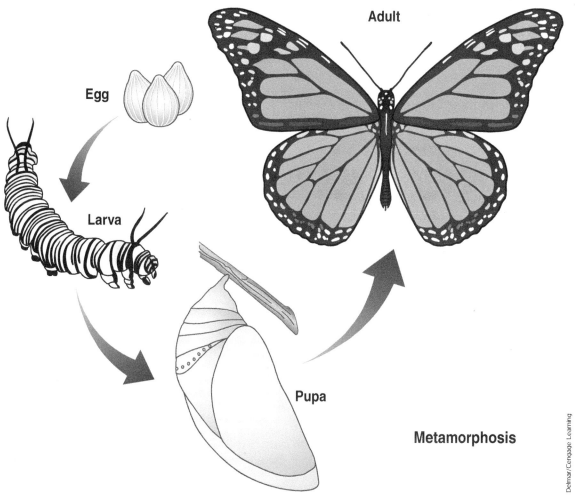

FIGURE 5-8

Insects go through a life cycle.

this stage because the larvae eat a tremendous amount of plant material. Larvae develop into the pupa stage and may be covered for a period by a cocoon. The pupa then develops into the adult stage. The adult lays eggs and the cycle begins all over again. Depending on the type of insect, they are more vulnerable at one phase than others. For example, if they spend the winter as a pupa in crop residue or in dead fruit tree limbs, getting rid of the crop residue or burning the dead limbs will get rid of the insects.

In integrated pest management, a combination of methods is used to control the insects. The crop is closely monitored and chemicals are used only when necessary. This not only cuts down on production costs but also helps protect the environment.

CAREER DEVELOPMENT EVENTS

Agriculture students should be aware of the damage that insects, diseases, and weeds create in all areas of agriculture. The combination of classroom study and FFA activities prepares students to handle pest problems correctly and safely. Agriculture teachers educate students on methods of identification, control, and prevention of these pests. The FFA offers students the opportunity to compete and win awards for their knowledge of this topic.

The National FFA Foundation provides funds for recognition of members' achievements at the local, state, regional, and national levels. Awards to members are generally presented at the annual chapter banquet. Competition includes team events and individual efforts to identify plant pests and disorders in forestry, floriculture, and nursery/landscape career development events. When students judge plants, they need to be able to look for symptoms of insect, parasite, and disease damage and presence.

FFA members may use a pest-related topic such as "Integrated Pest Management in Agriculture" to compete for the FFA public speaking career development events. Speaking out on agriculture issues not only develops skills in communication but informs the public about changing times. There are two FFA public speaking contests—a six- to eight-minute prepared speech and an extemporaneous speech.

Weeds

Weeds are plants that are growing where they are not wanted. If oats are growing in a wheat field, the oats may be considered to be weeds. In fact, some of our worst weed pests were once considered to be good plants. One of the best examples is Johnsongrass, which was brought in from Africa. In the 1800s, it was thought that this grass would be the perfect pasture grass in the South. It is a fast-growing grass that can be used for grazing. However, the grass got into other fields and competed with crops. It has now spread all across the country and is extremely difficult to kill in

AGRICULTURAL PESTS

In several FFA career development events, students must inspect plants for insect damage. Courtesy of Carol Duval, National FFA Organization.

FFA members have the opportunity to research and write prepared speeches. The speech manuscript is submitted to a panel of judges and used as part of the judging. The extemporaneous speech competition is designed to test the FFA member's ability to talk about a given subject without having prepared or rehearsed its content. The FFA member is given 30 minutes to generate a four- to six-minute speech on a given agriculture topic. Both events judge the student's power of expression, stage presence, voice control, knowledge, and response to questions.

crops. It outgrows most agricultural plants and robs them of sunlight and nutrients (Figure 5–9).

Weeds are pests in many different ways. They can harbor insects and diseases. Remember that insects go through several life phases. They may need weeds as a means of getting through one of the phases. Weeds compete with agricultural plants for nutrients and water. Weeds must have nutrients, the same as agricultural plants. If a field is full of weeds, the weeds may get as much or more of the soil's nutrients as the good plants. This means there are fewer nutrients for the crop. Also, the weeds may block out the sunlight (Figure 5–10). Sunlight is needed by the plants to produce sugars in the process of photosynthesis.

FIGURE 5-9
In the 1800s, Johnsongrass was thought to be a good pasture grass. Now it is a serious pest.

FIGURE 5-10
Weeds can block out sunlight from reaching other plants.

Another problem with weeds is that they may lower the quality of the crop. Hay that has weeds in it may be less nutritious or may even contain poisonous weeds. Grain that contains weed seed is not as good for human food or animal feed. A cotton crop that has weeds will have foreign particles in the lint and will be less valuable.

Most weeds are hardy, produce many seeds, and can completely take over an area very quickly (Figure 5-11). The fact that weeds are still around indicates that they are hard to control. People have been fighting them for thousands of years and the fight goes on.

People have used many means to fight weeds. Pulling the weeds up, hoeing, and plowing are the

FIGURE 5-11
Weed plants such as this Russian thistle usually produce a lot of seed. The wind will disperse the seeds over a large area.

oldest means. Also, animals have been used to get rid of unwanted plants. The animals are brought in and confined to an area where they eat the weeds. The most efficient way of controlling weeds is the use of a herbicide, a substance that kills plants. Chemical herbicides have been widely used since the 1960s and continue to be a very effective means of weed control.

There are basically two types of herbicides. One is the nonselective herbicides. This type kills all plants that it contacts. These are used in places where all of the vegetation is considered to be weeds. For example, nonselective herbicides are used on railroad beds to kill all plants. They also can be used with crops if the herbicide only contacts the weeds and not the crops.

The other type is selective herbicides. These will kill only certain types of plants (Figure 5–12). For example, a selective herbicide might kill cockleburs that are growing in corn. The herbicide will not affect the corn plants even if it comes into contact with them.

Years of research have gone into developing selective herbicides. A new herbicide must satisfy several requirements. First, it must not harm the crop it is used on. Second, the chemical has to be safe to use by producers. Producers come into contact with the herbicides as they are applied. If the herbicide is toxic to humans, the applicator

FIGURE 5-12

Selective herbicides kill only the weeds.

FIGURE 5-13

Pesticides have to be safe to apply.

might be harmed. Third, it must not be harmful to the environment. We have to protect plants and animals around us, so we have to be careful that the herbicide kills only the plants we intend. Fourth, the chemical must be affordable to use. It would do no good to use a chemical that is more expensive to use than the benefit produced. Modern herbicides must fit all of these requirements before they are released for use (Figure 5-13).

Some of the newer means of controlling weeds include the use of biological agents. This means that organisms such as insects and disease germs can be used. If an insect's favorite food is a particular weed, the insect might be used to control the weed. Of course, researchers have to be careful that the insect does not develop an appetite for crops, too. If the insect meets the requirement of only eating the weed pest, it can be raised for the purpose of eating the weed.

Diseases also can be used to fight weeds. Some diseases are particular about the type of plant they attack. Generally, a particular bacteria, virus, or fungus will attack only a certain type of plant (Figure 5–14). If a disease-causing agent can be found that only attacks a particular weed, the disease might be effective in controlling the weed.

Genetic Engineering

One of the most exciting possibilities for controlling agricultural pests is the use of genetic engineering. Genes are the substances in the cells of plants and animals that contain the material (DNA) responsible for passing characteristics from one generation to the next. Scientists now have the ability to take genes from one organism and place them in another. This means that they can alter the characteristics of plants and animals. Plants have been developed that are resistant to insects. This means that insects will not be as likely to feed on the genetically altered plants. Crops known as Bt crops are now common. These crops, such as corn and cotton, have been genetically engineered to produce a toxin from a soil bacteria, *Bacillus thuringiensis* (Bt). This toxin kills insects but is not harmful to humans or other mammals.

We now have the ability to make plants more tolerant of herbicides. Obviously, the more tolerant a crop plant is of a herbicide, the more likely weeds can be controlled. One of the problems is that the more similar the crop is to the weed, the more difficult it is to use a selective herbicide. By using genetic engineering, we can make the crop more unlike the weed.

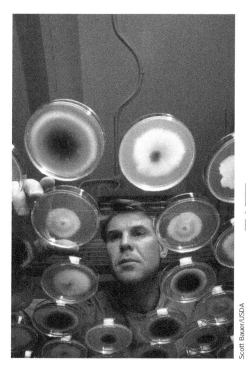

FIGURE 5–14
This scientist is studying a fungus that could be used to destroy weeds by giving them a disease.

SUMMARY

Humans have always faced pests that threaten agriculture. Although we spend a lot of time and money growing crops and animals, they are still a part of nature. In nature, plants and animals depend on each other. It is a difficult task to keep agricultural plants and animals away from the other organisms in nature. Through study and experimentation, we will better understand agricultural pests and find new and better ways of controlling them.

CHAPTER REVIEW

Student Learning Activities

1. Collect 10 insects that are harmful and 10 that are not harmful. Identify the insects and tell what damage the harmful insects do.
2. Identify a harmful insect that is common to your area. Devise a means of controlling the insect. Make sure your plan uses the IPM concept.
3. Identify the weeds that are common to your area. Collect five and bring samples to class. Discuss with the class how the weeds spread.
4. Locate plants that have a disease. Try to identify the disease.

True/False

1. _____ Insects do not carry diseases.
2. _____ Agriculture deals with the growing and cultivating of living things for human use.
3. _____ One of the basic tasks in caring for crops and animals is protecting them from other living things.
4. _____ Cancer and diabetes are diseases spread by mosquitoes.
5. _____ Some insects are parasites.
6. _____ Diseases also can be used to fight weeds.
7. _____ One of the biggest problems about insects is that most are deadly to humans.
8. _____ Insecticides are the most efficient means ever devised to kill insects.
9. _____ Any sound insect control program involves integrated pest management.
10. _____ Weeds are plants that are growing where they are not wanted.
11. _____ Genetic engineering is used in insect control.
12. _____ Almost all insects are harmful.

Multiple Choice

1. The insect life cycle begins when the female
   a. dies
   b. mates
   c. lays eggs

2. Insect eggs hatch into wormlike organisms called
   a. larvae
   b. pupas
   c. cocoons

3. A substance that kills plants is called a
   a. pesticide
   b. herbicide
   c. insecticide

4. Organisms that live on other organisms and harm them are called
   a. insects
   b. parasites
   c. predators

5. Many people have contracted diseases because of
   a. insects
   b. plants
   c. predators

6. An estimation of the number of major insect pest species in the United States is
   a. 750,000
   b. 2,000,000
   c. 500

7. Two insects that devour huge numbers of insects are
   a. lady beetles and praying mantises
   b. ticks and fleas
   c. cockleburs and honeybees

8. Johnsongrass is considered to be a
   a. landscaping grass
   b. perfect pasture grass
   c. weed

9. Plants have been developed that are resistant to
   a. insects
   b. pesticides
   c. animals

10. Every year cropland is prepared for
    a. animals
    b. insects
    c. planting

✳ Discussion

1. List and discuss some of the problems caused by insects.
2. Define integrated pest management.
3. What is meant by the biological control of insects?
4. Why are weeds considered pests?
5. What is the difference between a nonselective and a selective herbicide?
6. How can genetic engineering be used for controlling agricultural pests?
7. How do cultural practices have an impact on diseases and insects?
8. What is the life cycle of an insect?
9. How can weeds be controlled?
10. List two types of biological agents.

CHAPTER 6
Floriculture

Student Objectives

When you have finished studying this chapter, you should be able to:

- Define floriculture.
- Describe the international aspects of floriculture.
- Describe the national aspects of floriculture.
- Describe the three basic types of living plants florists sell.
- Describe the steps in producing a typical flower crop.
- List and describe the four flower-arranging principles.

Key Terms

floriculture
bulbs
florists
potted plant

foliage plants
houseplants
cut flowers
photoperiodic

The word floriculture literally means the growing of flowers. Floriculture is a very big business in the United States and the world (Figure 6–1). Not only are billions of flowers produced each day, but also they must be taken to their destination, arranged, stored, and sold. Each part of this chain is encompassed by the floriculture business.

International Aspect

Flower crops have been grown for centuries in Europe. Throughout history, flower gardens have produced beauty. Almost everyone enjoyed growing or looking at the flowers. Today, huge flower markets sell flowers by lots of hundreds and thousands. People seem to enjoy flowers as much as ever.

Most of the flowers brought into the United States come from Central and South America. The climate is favorable to the production of flower crops, and countries in the Southern Hemisphere can grow flowers during our winter. These are then shipped to the United States. Also, the cost of labor is low compared to other areas.

Other countries also send flowers to the United States. Australia and New Zealand sell their orchids here. Holland has always been famous for brilliantly colored tulips. Israel has begun producing roses and is rapidly becoming a leading source for these flowers.

Bulb Production

Another important component of the cut flower industry is that of producing bulbs. Just think of all the acreage of flowers grown worldwide and consider that somewhere bulbs have to be grown in order to produce the flowers. Worldwide there are over 64,000 acres of bulbs under cultivation (Figure 6–2). In the United States we produce almost 1,000 acres of flower bulbs. The worldwide value of these bulbs each year is over $2 billion. In the United States the value of flower bulbs produced is about $170 million.

FIGURE 6–1
Floriculture is a big agricultural business.

FIGURE 6-2
Producing bulbs is a major part of the floriculture industry.

By far, the largest production is that of tulips and lilies. Not only are these bulbs sold to commercial producers, but many are sold to consumers to plant in their gardens and around their houses.

In return, the United States ships gladiolus all over the world. The fresh flower industry is a worldwide business. Many countries have close ties to each other in the floriculture business.

National Aspect

As with the international aspect, flower production in the United States is concentrated in areas where the climate is best suited. Most floral crops need very bright light to grow and produce quickly. Also, some flowers need a cooler climate and others need a warm climate.

The most popular states for growing flowers are Florida and California because of the warm climates. Colorado and the New England states are also heavily involved in producing flowers that require cooler climates. Bulbs, such as tulips and daffodils, need slightly different production climates. Cool summers and mild winters make Washington, Oregon, and California more suitable for producing these (Figure 6-3).

FIGURE 6-3
Some flowers need a cool climate, and others need a warm climate.

FIGURE 6-4
Creating floral arrangements is a growing business.

Plants for Florists

Many different kinds of plants are produced for florists. Everything from blooming plants in pots and hanging baskets to cut flowers such as roses and daisies are available to florists. This is a growing industry in this country. People buy flowers for weddings, funerals, anniversaries, birthdays, and other occasions. People also send flowers to cheer up friends who are sick or just to let them know they are special (Figure 6-4).

Potted Plants

Any plant grown and sold in a pot is referred to as a potted plant. Outside plants such as azaleas and chrysanthemums are grown in pots for sale at the florists. These are usually removed from the pot and transplanted outside. Blooming houseplants like kalanchoe and cactus are sold in pots and generally remain in the pots (Figure 6-5).

Potted plants offer a challenge to the florist. Not only must they have beautiful or unusual foliage or blooms, but the plants have to be healthy. A potted plant, in most cases, must live, bloom, and remain attractive for a long time.

Most potted plants are grown from cuttings, while others are started from seed. It can take several

FIGURE 6-5
Blooming houseplants are sold and live in pots.

FIGURE 6-6
Foliage plants are grown for their attractive leaves.

weeks to a few months to produce a healthy, blooming potted plant.

Foliage Plants

Plants grown for their leaves are called foliage plants (Figure 6-6). Many times these are grown in pots and can also be called potted plants. Some plants, however, are grown for foliage for flower arrangements.

The most common plant for this use is the leatherleaf fern. Its fronds (leaves) are particularly strong. In many arrangements, leatherleaf accounts for most of the plant material. It is used to hide the support structure of the arrangement and as a green backdrop for the colorful flowers.

Another kind of fern, the Boston fern, is popular for hanging baskets. Other plants often used for hanging baskets are begonias and impatiens.

Often several small foliage plants are put together in shallow pots and sold as dish gardens. Groupings of plants inside glass or plastic containers are called terrariums (Figure 6-7).

The most popular foliage plants, however, are those grown for homes and public buildings. These can be sold in small pots but are also offered in very large sizes. Landscapers use these to make the yards and outsides of homes, businesses, and other structures more attractive.

FIGURE 6–7
Plants can be used in a variety of ways to create beauty.

Exotic Plants

Unusual or rare plants are called exotic plants. Many exotic plants are great for houseplants. These are attractive because they are so unusual.

An example of an exotic plant is the bromeliads. These plants are unusual because of their cup-shaped leaves. The leaves act as cups because they hold water. For this reason, they can stand longer periods of neglect. Another example is the carnivorous plants. These are plants that eat insects to grow. The Venus flytrap is the most common carnivorous plant. It traps flies and other insects and digests them (Figure 6–8). Another carnivorous plant is the pitcher plant. It traps insects by luring them into its "throat." The sides are so slippery the insect cannot get out. It then falls into the digestive fluids in the bottom of the "pitcher" and dies. There is a good market for these exotic plants. People buy them to decorate their homes.

Cut-Flower Production

The main items in the floriculture trade, of course, are flowers. Cut flowers are flowers that are to be used in flower arrangements. They are cut at the

FIGURE 6-8
The Venus flytrap is an exotic plant that catches insects.

FIGURE 6-9
Cut flowers are trimmed so they can fit into a vase.

stems so they can fit into vases and other containers (Figure 6-9). This differentiates cut flowers from flowering plants. Flowering plants may be sold as potted plants. The plants that produce cut flowers are not sold. Only the flowers and stems are marketed.

Many nurseries produce flowers just for florists. Many produce only one kind of flower, such as chrysanthemums or carnations. For the growers to make a profit from their crops, they must practice good growing habits and timing.

Physical Environment

Although some cut flowers, such as gladiolus, are often grown outside in a field, most are grown in greenhouses. There are several reasons for this.

First is temperature. To induce plants to form buds, then flowers, they must have a certain range in temperature. Cold temperatures will keep many plants from forming buds at all. High temperatures may cause them to abort buds and blooms. Temperature can be controlled inside a greenhouse.

Another important factor is moisture. Plants for cut-flower production must have just the right amount of water to do their best. Producers can use sprinklers or underground water to keep plants in the field watered. However, it is difficult to prevent the plants from getting too much rain. Hard rain can also destroy the fragile blooms. Just as rainstorms can damage plants, so can windstorms. By growing the plants in the controlled environment of a greenhouse, these conditions can be avoided (Figure 6–10).

CAREER DEVELOPMENT EVENTS

The FFA Floriculture Career Development Event is a competition for teams of four members. Each member competes individually and the final team score is the total of the three highest individual scores. The event consists of four parts: (1) identification of plant materials, (2) placing classes, (3) general knowledge, and (4) practicum.

The identification of plant materials consists of plants selected from the State Floriculture list and is displayed for visual identification. Students may not touch the live plant material but can use a magnifying glass to inspect the foliage. The judging classes are plant materials or floral arrangements. The student places the materials from best to worst, based on structural form, market value, and health. Floriculture plants are selected from the identification list and may be cut flowers, foliage plants, bedding plants, or potted plants. The general knowledge part of the floriculture event consists of 50 multiple-choice questions. This phase of the event tests knowledge and understanding of all areas of the floriculture industry. Students preparing for this part of the event must study not only the operations and maintenance of a greenhouse, but also the physiology, growth habits, nutritional needs, disorders, and propagation practices of floriculture crops. Questions addressing floral design, arrangements, equipment, and principles are included as well as customer service, interpersonal relations, salesmanship, and records and reports.

The part of the event known as the practicum is where the student identifies plant disorders. Plant disorders include disease, insects, weeds, and nutritional problems. The disorders are displayed for the student to view with or without a magnifying glass. The student is asked to identify the problem and those pesticides or other treatments necessary to remedy the problem. Hands-on practicums are included in the event. Students are asked to perform different skills required in the floriculture industry, such as potting plants, designing a floral arrangement, or taking a phone order.

Greenhouse Benches

Benches in greenhouses used for growing potted plants are usually from 5 to 8 feet wide. Benches used for the growing of cut flowers are only 4 feet wide. This is helpful for the intense care these plants need (Figure 6–11). Plants used for cut flowers are usually

The Floriculture Career Development Event tests students' ability to work with floral arrangements. *Credit Carol Duval, National FFA Organization.*

After all team members have completed each phase of the event, each individual score is calculated. The top three individual scores are added to yield the team score. If there is a tie between two team scores, the written exam score is used as a tiebreaker for first- and second-place teams. Winning teams from each state advance to the National Floriculture Career Development Event.

The FFA Floriculture Proficiency Award is one of the four horticulture proficiency awards. The award is presented to individual FFA members who have excelled in experiences using the principles and practices of field or greenhouse production of flowers, foliage, and related plant materials for ornamental purposes. To apply, the student must be involved in a Supervised Agricultural Experience (SAE) and have maintained a list of floriculture skills he or she has learned. The student must also develop an inventory of buildings, equipment, tools, and materials used in the experience program. He or she must also keep accurate records of income, expenses, and accomplishments. Students may compete within their chapter, district, and state. The state award winners compete for national recognition.

FIGURE 6-10

In a greenhouse, flower growers can control the environment.

FIGURE 6-11

Greenhouse benches hold pots for growing plants.

not grown in pots. Instead, the benches hold soil and the plants are grown in this. These benches are also closer to the ground. Often, flower stalks must be staked or tied to produce straight stems, especially with the large flowers. Lower benches allow more room in which to work.

Growing Cut Flowers

The grower usually starts with a cutting. Let's follow a typical crop of chrysanthemums. Some growers make their own cuttings, root them, grow them out, induce blooming, harvest the flowers, and ship them. More often, they start with rooted cuttings (Figure 6-12).

FLORICULTURE

FIGURE 6–12
Some growers start new plants by taking cuttings as shown in (A) and doing rootings as shown in (B).

This is where timing comes in. Growers usually have several crops growing at once so they can have continuous production. A new crop is planted every two to four weeks. Different types of chrysanthemums mature at different times, so the grower must take this into account also.

The cuttings are planted in the low, narrow benches and grown out. Wire mesh, usually with 6-by-8-inch spacing, is laid on top of the soil before planting. This is raised as the plants grow to ensure straight stems on the flowers.

About two weeks after planting the plants are pinched. This refers to the practice of taking off the bud at the very tip of the plant. The plant will then produce several stems from buds farther down the stem (Figure 6–13). All but two or three of these stems are taken off. This makes the plants produce more flowers.

Lighting

Many plants such as chrysanthemums are **photoperiodic.** This means that they respond to the length of night. Only during long nights, or short days, do chrysanthemums produce flower buds. Left alone, they flower only during the fall and early winter. To get around this, the dark period is artificially

FIGURE 6–13
Pinching makes the plants produce more limbs.

controlled. To allow the plants to grow for a while without forming buds, the dark period is interrupted with electric lights.

When the plants are larger and the stems are long enough, lighting is no longer as important. If the greenhouse is in a rural area, without outside light coming in and it is the right time of year, the plants will form flower buds naturally. Most of the time, however, this is not left to chance. The plants are covered each evening with a solid piece of black cloth that blocks all light (Figure 6–14). The cloth is then removed in the morning. It only takes a very small amount of light from almost any source to interrupt the cycle. This continues until the flowers are well developed.

Temperature

Temperature must be monitored frequently. Although different cut-flower plants need slightly different temperatures, chrysanthemums need a night temperature of 60 to 62 °F and a day temperature of 65 to 70 °F.

If temperatures go above 80 °F or so, it takes longer for flowers to be produced and the flower quality is lowered. By contrast, if temperatures are too cool, chances of disease go up, production time is longer, and the flower size and stem length are both less.

FIGURE 6–14

A cloth over the greenhouse can help control the amount of light the plants get.

Timing

The grower must follow a precise timetable. This timetable must be started from the date the flowers must be ready and worked backward to pinpoint when the plants should be started, when short or long days should be started, when the plants should be pinched, when fertilized, and when temperatures should be controlled.

Timing is more important with some flowers. For example, unlike chrysanthemums, which are used all year in flower arrangements, poinsettias are grown for and bought only at Christmas. Green poinsettias are unsalable in December. Poinsettias that are a beautiful red and green on December 26 are likewise useless. Rigid schedules of lighting, watering, and temperature control are used to make them bloom at the right time (Figure 6-15).

Harvesting

Chrysanthemum flowers may be harvested when they are just opening or in full bloom. This depends on the type of flower and how far the flowers will travel to market. The stem is cut as long as it can be. This is to give the florist enough stem length for a variety of uses. Harvesting continues for one to two weeks. Then the plants are discarded and the bed is prepared for the next crop.

FIGURE 6-15

Poinsettias require rigid schedules of lighting, watering, and temperature control.

FIGURE 6-16
The best grade of flowers are large and perfect.

The flowers are immediately put into room-temperature water, then into a cooler where they are kept at a much-lower temperature. Once the flowers are cut, they are no longer living and will begin to decay. Cool temperatures delay the decay until the flowers are used. Later, the flowers are graded. The best grade includes the largest, most perfect flowers (Figure 6-16). Smaller flowers or slightly off-color flowers are lower grades.

Transportation

Flowers are easily damaged and special measures are taken to protect them (Figure 6-17). The bunches (for

FIGURE 6-17
Flowers have to be carefully packed to prevent damage.

chrysanthemums there are usually 12 in a bunch) are wrapped in a clear plastic. Large, sturdy cardboard boxes are used to further protect the flowers. Newspaper is packed gently around the flowers to keep them from jostling.

Flowers are transported in refrigerated trucks for short distances. For longer distances, such as from California to Maine, or Colombia to Miami, they are shipped in refrigerated compartments in airplanes.

Some of the more delicate flowers only last a matter of days, then are unusable. For this reason, the florist must know when to order these.

Flower Arrangements

How well the flowers look depends to a large degree on how they are presented or arranged. Among the principles of flower arranging are balance, scale, focal point, and unity (Figure 6–18).

Balance refers to the apparent weight of the flowers. An arrangement should not look like it is about to fall over. By balancing large flowers and light and dark flowers, the designer creates a beautiful piece that looks stable.

Scale involves the various sizes of flowers compared to each other and the container. It would not look right if the arrangement consisted of a few small flowers in a very large vase.

An arrangement should have something that draws the viewer's attention. This is called the focal point. The focal point is usually in the front, near the bottom of the arrangement. It can be a very large flower, a very dark flower, or an unusual flower.

The design should flow together. A wildflower bouquet with one rose would be out of place. Unity is necessary in an arrangement. All the parts should fit as one.

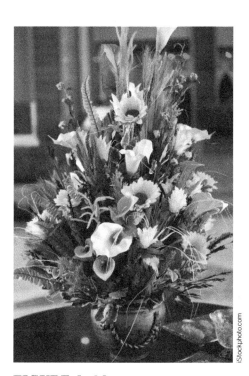

FIGURE 6–18
A nice floral arrangement must have a focal point, balance, scale, and unity.

SUMMARY

From growing flowers and plants to selling the finished product, floriculture is a worldwide business. It takes planning and forethought to produce and sell floriculture products.

CHAPTER REVIEW

Student Learning Activities

1. Bring a variety of different types of flowers to class. Arrange them in different ways. Try to identify several principles that apply to making floral arrangements that look good.
2. Interview a florist. Determine where the flowers used by the florist come from and the main markets for the arranged flowers. Share your findings with the class.
3. Obtain a potted plant from a department store, florist, or friend. Go to the library and conduct research about the plant. Find out the country where the plant originated and where the domesticated plants are grown. Also determine how the plants are produced. Report to the class.
4. Choose a type of flower, such as roses, tulips, poinsettias, orchids. Research the history of the flower. How far back in time have humans cultivated them? What countries developed them? What interesting facts can you discover?

True/False

1. _____ Most potted plants are started from seed.
2. _____ Floriculture is the growing of flowers.
3. _____ Plants grown for their leaves are called foliage plants.
4. _____ Photoperiodic means that the plants respond to the length of the day.
5. _____ If temperatures rise above approximately 80 °F, it takes longer for flowers to be produced and the flower quality is lowered.
6. _____ Balance refers to the apparent weight of the flowers in an arrangement.
7. _____ The most common plant grown for foliage is the pitcher plant.

8. _____ Dish gardens are deep pots for small foliage plants.

9. _____ Unusual or rare plants are called exotic plants.

10. _____ Terrariums are plants growing inside a glass or plastic container.

Multiple Choice

1. Most of the flowers brought into the United States come from
    a. China and Japan
    b. Central and South America
    c. Australia and New Zealand

2. One of the leading sources for roses is
    a. China
    b. Israel
    c. Australia

3. Flower production in the United States is greatest in
    a. Florida and California
    b. Georgia and Florida
    c. North Carolina and California

4. The main items in the floriculture trade are
    a. shrubs
    b. plants
    c. flowers

5. Benches in greenhouses used for growing potted plants are usually between
    a. 5 and 8 feet wide
    b. 3 and 4 feet wide
    c. 1 and 2 feet wide

6. The most obvious example of the importance of timing involves
    a. poinsettias
    b. chrysanthemums
    c. leatherleaf ferns

7. About two weeks after planting, the plants are
    a. graded
    b. pinched
    c. cut

8. Cut flowers are used for
    a. flower arrangements
    b. foliage
    c. hanging baskets

9. Holland is famous for its
    a. daisies
    b. roses
    c. tulips
10. Fronds are another name for
    a. buds
    b. stems
    c. leaves

✸ Discussion

1. Why are most cut flowers grown in greenhouses?
2. What does scale mean?
3. Where do most flowers bought in the United States come from and why?
4. Why are potted plants a challenge to florists?
5. How does a pitcher plant trap insects?
6. How can temperature, rain, and wind affect plants?
7. Why is floriculture such a big business throughout the world?
8. How are cut flowers grown?
9. Why are poinsettias such a good example of timing?
10. Give some examples of exotic plants.

CHAPTER 7
Nursery Production

Student Objectives

When you have finished studying this chapter, you should be able to:
- Define a plant nursery.
- Describe materials nurseries use.
- Identify the basic greenhouse types.
- Describe what a mist system does.
- Distinguish sexual from asexual propagation.
- Describe the importance of breeding programs.

Key Terms

plant nurseries
stock plants
media
bedding plants
cold frames
shade cloth
propagation
cuttings
division

Have you ever wondered where all the flowering plants, shrubs, and trees around our houses come from? They come from **plant nurseries** (Figure 7–1). People who work at nurseries start with seeds or pieces of plants and grow them out to whole plants. These plants are sold in everything from 1-inch pots to 50-gallon pots. Some are even grown directly in the ground, then dug up when sold. Some of these plants must be moved with huge trucks, due to their size.

Some growers produce their own plants, grow them out, then sell them. In some nurseries there are entire teams of people who only take cuttings. These people are very skilled and can take up to 8,000 cuttings in a single day.

After these cuttings are taken from the **stock plants** (plants grown for the purpose of producing cuttings), they are treated with a chemical that induces rooting, then placed in soil in greenhouses (Figure 7–2).

Many nursery businesses buy cuttings or small plants (called liners) and grow them to sell. The liners are potted up in larger containers and allowed to grow for a few months to a year. After the plants have filled out the larger container, they are either sold or transplanted into larger containers that will be sold later.

Plants bought from other sources for the grower to work with should arrive quickly and be healthy.

FIGURE 7–1

The plants around your home probably originated at a nursery.

FIGURE 7-2

Cuttings from superior plants are used to start new plants.

Unrooted cuttings do not survive long. They must be handled very carefully. Frequently they are put into a cooled room until they can be processed.

Growing Media

The substances that plants are grown in are called media. Why not call it soil? Because many plants are not grown in soil. Over the last several years growers have begun to rely more and more on "soilless" media.

Media must combine several characteristics. They must be well drained so the roots of a plant don't rot. But at the same time, they need the ability to retain water for the plant's use.

Soilless media are usually made of a mixture of ingredients (Figure 7–3). A basic component can be peat. Peat is compressed organic matter, usually harvested from bogs that have been drained. Another common ingredient is pine bark that has been crushed into small pieces. Sand is added in some instances.

Human-made, or altered, elements include vermiculite (which is actually heat-expanded rock), rock stuff, and perlite. These artificial ingredients have the advantage of being lightweight. This is especially important for nurseries that ship their plants using freight companies that charge by weight.

FIGURE 7-3

Soilless media are mixed from a number of ingredients. From left to right, these ingredients are peat, perlite, and vermiculite.

FIGURE 7-4

Seeds are usually started in flats.

Containers used in the nursery industry vary widely. Many seeds are started in flats (Figure 7-4). These are trays filled with media. Usually fine, lightweight media are used to make it easier for the seedlings to grow.

Bedding plants (plants, usually annuals, that are planted in large quantities) are usually sold in cell packs (Figure 7-5). Cell packs come in several sizes, from three packs to the new, larger jumbo packs. These are convenient for the person purchasing the plants. The plants are taken directly from the cells and planted.

NURSERY PRODUCTION

FIGURE 7-5
Bedding plants are planted and sold in nursery cell packs.

FIGURE 7-6
Large plants are dug up and the roots are wrapped in burlap. This is called being balled and burlapped.

Many plants are started in 2.5-inch pots. After growing in these pots, they are often transplanted into larger pots. Some are produced as liners to be sold to other nurseries.

For very large trees and shrubs, containers as large as 30 to 50 gallons are sometimes used. Large plants grown in the field are dug up and burlap is wrapped around the root system (Figure 7-6). This helps keep the roots moist and bound together to prevent injury. The burlap can then be planted with the tree. It decays in a short time.

Most growers use plastic pots, although a few use clay. Clay pots are much heavier than plastic ones and are easier to break. Clay pots are also porous, meaning they have tiny holes throughout them. This ensures that the plant roots will get oxygen and avoids overwatering problems. By contrast, because the clay pots are porous, water is lost faster and the plants must be watered more often (Figure 7-7).

The color of the pots is sometimes important. Black pots tend to absorb more of the sun's energy and keep the roots warm. This is good in winter, but

FIGURE 7-7

Plastic pots (A) are lighter and will not break. Clay pots (B) are porous and allow oxygen to get to the roots of plants.

(A)

(B)

it is sometimes a problem in summer. White pots keep the roots cooler, but they break down in the sunlight and do not last as long as black plastic pots. A few nurseries color-code their crops. Blue, pink, orange, and other colors may signify the time the plant was repotted, the type of plant, or the color of the blooms.

Growing Areas

Most nurseries have both inside and outside growing areas. Both the climate and the plants grown have an effect on the amount of outside and inside growing areas. Plants such as daylilies can be grown in fields.

Houseplants, such as poinsettias for Christmas, must be grown in covered, heated areas.

Greenhouses are the most common enclosed growing areas. These structures allow light in and protect plants from the weather. Greenhouses are usually heated. This allows the grower to extend the growing season and produce plants all year long. Moisture can be precisely controlled in greenhouses. Plants can be given water only when they need it. This protects the plants from overwatering during very wet weather.

Most modern greenhouses are made of fiberglass or polyethylene (Figure 7–8). Fiberglass is a rigid material and is attached to a metal or wood frame. Fairly inexpensive, fiberglass weighs less than glass but yellows with age, cutting down on the amount of sunlight that gets through. Polyethylene resembles large sheets of plastic and is even less expensive than fiberglass. However, it only lasts one to five years and then must be replaced. These large pieces of polyethylene are also attached to a metal or wooden frame.

The shape and size of the greenhouse used depends on what the nursery is producing and where it is located. In the North, the greenhouse must be able to withstand snow loads. In the South, it is more important that the greenhouse be easily cooled.

FIGURE 7-8
Most greenhouses are covered in fiberglass or plastic.

FIGURE 7–9

Many modern greenhouses are Quonset shaped.

Most of the modern nurseries are putting in Quonset greenhouses (Figure 7–9). These look like a barrel cut in half. Almost all of these are built to be covered in polyethylene, although a few are manufactured for fiberglass coverings.

Freestanding even- or uneven-span greenhouses are better able to take snow loads. Some of these structures are covered with glass, but most are covered with fiberglass.

In larger nurseries, several greenhouses are connected at the sides. This is called ridge and furrow. It allows for several different climates inside while still being more cost efficient as far as heating and cooling goes.

A group of greenhouses at a nursery is called a range. A range can be made of all one type of greenhouse, such as Quonsets, or a mixture of types, such as Quonsets and freestanding or even-span greenhouses.

Some greenhouses are put up just to shield plants from wind and very low temperatures. These greenhouses have no supplemental heat in them. Because of this, they are called cold frames (Figure 7–10). Cold frames can be small structures that are mobile and just laid over plants in a bed, or they can be very large with benches, just like heated greenhouses.

FIGURE 7-10
Cold frames have no artificial heat and only provide shelter for the plants.

Outside Growing Areas

The most common outside growing structure is the lath house. A lath house is a metal or, more commonly, a wooden structure with shade cloth stretched over the top and sometimes sides. Shade cloth is loosely woven black cloth that blocks part of the sun (Figure 7-11). These black cloths are commercially available in weaves that block from 5 to 90 percent of the sunlight.

FIGURE 7-11
Shade cloths partially block out the sun.

CAREER DEVELOPMENT EVENTS

The Nursery Operations Proficiency Award is an individual competition. FFA members who have job experience with turf, bedding plants, shrubs, and/or trees for production operations may apply for the Nursery Operations Proficiency Award. To apply, the student's SAE, school classes, and FFA activities should be related to this area. Isabel was the winner of the chapter proficiency award in nursery operations. She had spent over 1,800 hours working for a local grower. Isabel's work experience included propagating shrubs, pruning field-grown trees, fertilizing container stock, and applying chemicals and fertilizer. Isabel also developed skills in communication and computer applications. She was active in her FFA fund-raising and community awareness activities and served as chapter secretary for two years. Isabel's accurate record keeping and wide range of experience and activities made her an excellent candidate for a proficiency award. All proficiency awards are judged on the students' application and experience. Good record keeping plus the development of proficiency-related skills and accomplishments are essential to becoming a winner. Students may compete within their chapter, district, and state. The state winners compete for national recognition.

Another opportunity for individuals to be awarded through their interests in the nursery industry is the FFA achievement awards. Achievement awards are lists of skills, abilities, and attitudes needed to enter a career in the area of the student's interest. The Ornamental Horticulture Achievement Award is divided into four areas: Career Exploration, Career Skills, Leadership Development, and Safety Practices. Each of these four areas has a list of skills and abilities, and the student has to master 80 percent of each list to receive the award. The agriculture instructor or even a worksite employer can evaluate and check the student's accomplishments.

Eddie wanted to learn more about the nursery industry and set out to win the Ornamental Horticulture Achievement Award. Eddie first worked on accomplishing the Career Exploration skills by visiting and interviewing five different horticulture

NURSERY PRODUCTION

FFA members who enjoy working with plants can get involved with projects in nurseries and apply for the Nursery Operations Proficiency Award. *Courtesy of Frank Flanders*

operations and presenting a written report to his teacher. He also had to learn the skills required to compete as a team member in the Nursery/Landscape Career Development Event and participated in the FFA chapter tree-planting program. To satisfy the Career Skills, Eddie prepared a scrapbook. The scrapbook identified 100 plants, nursery media, pruning techniques, insect and disease problems of 20 landscape plants, and a list of tools and equipment needed in a nursery operation. Next, Eddie had to demonstrate to his teacher the proper ways to propagate a shrub, plant and prune a tree, apply fertilizer, spray insecticide, and keep records of his plants in the school greenhouse. Eddie obtained his points in Leadership Development by attending chapter meetings, participating in chapter fund-raising activities, serving as chapter chaplain, and entering the public speaking contest. Eddie's employer at the landscape business checked his knowledge of safety practices. Eddie demonstrated the safe use of tools and equipment used in his job. After Eddie completed 80 percent or more of the skills as verified by the evaluator, he was presented with the official FFA Achievement Award Certificate at the school's award program.

All FFA members should strive to earn at least one achievement award. After deciding on your career interest, work with your parents and advisor to select the skills and experiences you want to attain.

Unlike greenhouses, water is not controlled and temperature is only changed a little. These houses are used only to provide shade to those plants that do not like a lot of sun.

Some nurseries use outside beds to grow out plants. Treated boards are used to form raised beds. These are then filled with sand, soilless media, or plain soil. Plants are grown in them, then sold bare root, or without a pot. Sometimes these plants are dug up and then potted into containers for sale.

Some areas outside are covered with landscaping cloth. Landscaping cloth is a thick, black cloth that forms a barrier to weeds and holds down erosion. Containers are put directly on the landscaping cloth and grown out, then sold.

Another area that is used for growing plants outside is the ground itself. Plants are put into the ground, allowed to reach a good size, then harvested and sold bare root.

Mist Areas

Some plant reproduction techniques involve the use of mist. Tiny droplets of water from irrigation lines keep the plants moist so they don't use up their water reserves (Figure 7–12). There are two main types of mist systems used. The first type is a timed system.

FIGURE 7–12

Irrigation systems in modern greenhouses are timed to water at certain times.

A solenoid wired into the irrigation lines controls the flow of water. The solenoid is controlled by a timer. The water goes on for a short time (a few seconds) every five minutes to every hour, depending on circumstances.

The second type uses a mechanical leaf. A piece of wire is attached to an electrical device. When the screen (the mechanical leaf) is wet, the weight of the water on the screen tilts the screen down, trips the solenoid, and shuts off the water. As the screen dries out, it becomes lighter. When this happens, it tilts up and trips the solenoid to let water through. The water makes the mechanical leaf heavy again and turns the solenoid off.

Propagation

Plant reproduction, or making several plants from one plant, is called propagation. This is done several ways, depending on the plant and the nursery operation.

Some operations produce their own seed and grow plants from them. Others buy seed to begin plant production. Some only handle pieces of plants. Others may combine the two main propagation methods.

Sexual Propagation

Sexual propagation refers to producing plants from seeds. Plants have been producing seeds for many millions of years. Humans began taking advantage of this at the dawn of civilization. Plants that they found useful were propagated to produce new plants.

Sexual propagation begins with the flower. Flowers differ in size, color, and parts. Some plants have only female flowers on them, others only male. Other types of plants have male and female flowers separately on the same plant. Most plants, however, produce flowers that are both male and female.

The female parts of a flower, called the pistil, consist of the stigma, the style, and the ovary. Egg cells are produced in the ovary.

The male parts of the flower, called the stamen, contain the anther and filament. The anther produces pollen, or sperm cells. The pollen, carried by the wind or insects, lands on the stigma. This is called pollination. It then forms a pollen tube through the style to the ovary. When the sperm cell actually unites with the egg cell, fertilization has occurred.

Flowers fertilize each other or themselves to produce seeds. When they fertilize themselves, they are self-pollinating. When they pollinate another flower, they are cross-pollinating. When the flowers pollinate, the male and female cells are united. This forms the zygote that develops into the seed.

Many nurseries that produce annual bedding plants start with seed they order from another nursery (Figure 7–13). They germinate the seed, then grow it out. Germination is when a seed starts to grow into a plant.

Growing plants from seed is less expensive than growing plants from cuttings. It enables a grower to produce many thousands of plants quickly.

Asexual Propagation

Sometimes it is less desirable to start plants from seed, especially for large trees and shrubs. It takes longer to get a salable plant starting from seed. In some instances, a plant doesn't "come true" from seed. That means that the plant from a seed may not

FIGURE 7–13

Annual plants are often started from seed.

NURSERY PRODUCTION

be exactly like the parent plant. This is because of cross-pollination, where characteristics are obtained from both parent plants.

For this reason, many growers produce their plants using asexual propagation. This method starts with a part of a plant itself, instead of a seed produced by the plant.

Cuttings

The most popular form of asexual propagation is taking **cuttings** from the plant. A piece of the stem (or sometimes a leaf) is taken from the plant. It is usually treated with a rooting hormone (a chemical that makes the plant form roots), then placed in media (Figure 7–14). The cutting is then placed under mist until it has formed roots and no longer needs the extra water.

Propagating plants this way assures the grower of a crop that will be almost exactly like the original. In addition, the plants will be uniform in growth. This is especially important for landscapers who buy many plants for hedges or beds.

Some plants cannot be propagated this simply. In this case, other asexual methods are used. Some plants are propagated by air layering. The stem is peeled all the way around, a rooting hormone is painted on the wound, and the area is then wrapped in damp media, covered with plastic, and

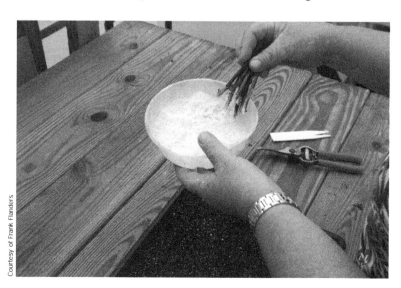

FIGURE 7–14

Root hormones are applied to help a cutting grow new roots.

FIGURE 7-15
Many plants can be propagated by separating or dividing the roots or bulbs. These Dahlia bulbs will be separated and produce several new plants.

allowed to form roots while still getting water from the parent plant.

Other plants must be propagated by root cuttings. The roots are cut off much as the stems are, then allowed to produce a plant.

Tissue Culture

Micropropagation, or tissue culture, produces an entire plant from a single cell or group of cells of the parent plant. Although it takes quite a bit longer than producing plants from cuttings, it has the potential to produce millions of plants from a single branch of a plant.

Still other plants must rely on **division,** or separation, for propagation (Figure 7-15). This involves digging up part of a plant just below the soil surface and separating it into several parts. Some plants can be gently pulled apart, called separating. Other plants must be cut apart into sections; this is called dividing. These parts are then allowed to grow into whole plants.

 ## Plant Production

Let's follow a typical plant from establishment to selling. One of the most popular plants sold in nurseries today is the holly. Hollies are valued for their glossy green foliage, red berries, and screening ability (Figure 7-16). A row of hollies can screen a busy highway or noisy neighbors. They are common in landscaping plans.

Propagating *Ilex crenata*, the Japanese holly, begins with stock plants. These are large plants that are either planted on the nursery grounds to be used as cutting material, or older plants in large containers.

The cuttings must be taken from the right part of the plant. Branches that look sick or were heavily shaded by the rest of the plant should be avoided. Old branches with thick bark should also be left. Very young, weak tips may rot under mist. Usually young, firm wood is taken. The cuttings are between 4 and 6 inches long.

NURSERY PRODUCTION

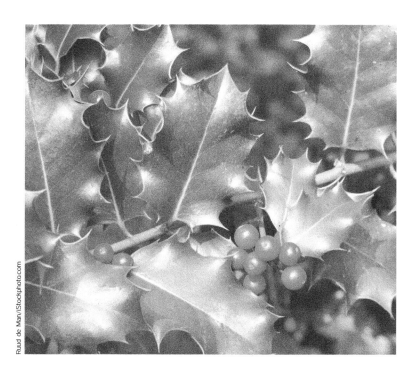

FIGURE 7-16
Holly is valued for its bright leaves and berries.

The best time to take cuttings is in the cool morning. If exposed to heat and direct sunlight, the cuttings may suffer enough damage to be unusable. Cuttings are kept in a cool, shaded place and kept moist until ready to process. Depending on the operation, a few hundred or several thousand cuttings may be taken before they are all processed.

The lower leaves are taken off to expose the bottom node. Often the bark is scraped off in a small area to help the plant to form roots. A powdered or liquid rooting hormone is then applied to the bottom of the stem.

The cuttings are then stuck into moist media in small pots, bare root, or outside beds under mist. Keeping the new cuttings moist is especially important. If allowed to dry out too much, the cuttings will die. If the area is kept too wet, fungus and bacteria may become a problem.

The cuttings stay there until they form roots. They are misted less and less frequently. After three to four weeks, the cuttings are weaned off the mist.

Once the root system has filled the small pots, they are repotted into 1-gallon containers or sold as liners. When the roots have filled the 1-gallon pots,

FIGURE 7–17
Young holly plants are sold in pots.

they are either sold or repotted again. Hollies are available in as large as 50-gallon pots. This size is rather expensive, however, and the majority of hollies are sold in 1- to 3-gallon pots (Figure 7–17).

When growing out, the hollies are fertilized, either with constant liquid fertilizer, or a granular fertilizer put on one to three times a year. They are sprayed for pests, and the pots kept weed free. Water is used when needed, usually twice a week or so, depending on weather conditions.

SUMMARY

There are many things to consider when thinking about the nursery business. Nurseries vary widely in the plants they grow and the methods they use. The nursery business is complex and fascinating.

CHAPTER REVIEW

Student Learning Activities

1. Take an inventory of the plants around your home or school. Determine which of these originated in a nursery.
2. Visit a nursery in your area. Find out which plants sell the most. Think of some reasons why these particular plants are the most popular. Report to the class.

3. With your teacher's assistance, take cuttings from a plant and root them. Grow the plant in a pot and at the proper time plant it at your home.

4. Examine the plants around your school. Identify each and determine how each was propagated.

True/False

1. _____ Plants are the basis of the nursery industry.
2. _____ Unrooted cuttings survive a long time.
3. _____ Vermiculite is a human-made element.
4. _____ The most common outside growing structure is a greenhouse.
5. _____ Landscaping cloth is used solely to keep insects away from plants.
6. _____ The most popular form of asexual propagation is taking cuttings from plants.
7. _____ Micropropagation produces an entire plant from a single cell of the parent plant.
8. _____ The best time to take cuttings is in the afternoon.
9. _____ Stock plants are grown for the sole purpose of producing cuttings.
10. _____ All substances that plants are grown in are called media.

Multiple Choice

1. Human-made elements include
    a. peat
    b. bark
    c. perlite

2. The most common enclosed growing area is the
    a. nursery
    b. greenhouse
    c. lath house

3. Most modern greenhouses are made of
    a. plastic
    b. metal or wood
    c. fiberglass or polyethylene

4. Greenhouses put up to shield plants from wind and temperatures are called
    a. cold frames
    b. ranges
    c. lath houses

5. Shade cloth is used to
   a. kill weeds
   b. block out the sun
   c. hold down erosion

6. Sexual propagation refers to producing plants from
   a. seeds
   b. leaves
   c. cuttings

7. Egg cells are produced in the
   a. ovary
   b. pollen tube
   c. stigma

8. When flowers pollinate another flower, it is called
   a. self-pollination
   b. cross-pollination
   c. fertilization

9. Freestanding even- or uneven-span greenhouses are better able to take
   a. water loads
   b. soil loads
   c. snow loads

10. The most common outside growing structure is the
    a. greenhouse
    b. lath house
    c. range

✳ Discussion

1. Why are plastic pots more common than clay pots?
2. Why is the color of plant pots important?
3. What are the two main types of mist systems?
4. What is the difference between sexual and asexual propagation?
5. List the male and female parts of a flower.
6. What is air layering?
7. What are media?
8. What is a lath house?
9. What is the difference between self-pollination and cross-pollination?
10. When is the best time to take cuttings? Why?

CHAPTER 8
Landscaping

Student Objectives

When you have finished studying this chapter, you should be able to:

- Define the term landscape.
- Describe the duties of a landscape architect and a landscape designer.
- Describe the different types of plant materials most commonly used by landscape architects and designers.
- List and describe the phases of landscape development and construction.
- List and describe the five principles of landscape design.
- Describe the duties of a landscape contractor.
- Define the term interiorscaping.

Key Terms

landscaping
landscape architect
landscape designer
landscape contractor
emphasis (focal point)
balance
proportion and scale
rhythm
variety and unity
topiary
interiorscape

Designing the Landscape

What do you think of when you hear or read the word "landscape"? Most people imagine mountains, water, forests, or wild meadows. The term landscape actually means a view of any natural scenery as seen from one perspective (Figure 8–1). **Landscaping** is the act of taking a certain plot of ground, usually in its natural state, and redesigning it to be more comfortable, attractive, and useful for humans or other animals. For example, your local park has most likely been landscaped.

If you have a backyard, it probably has been landscaped, too. Even the habitat areas in zoos are a result of careful planning so that the animals feel that they are in their natural homes. Some people like the look of natural areas such as forests or wild meadows, whereas others prefer to arrange landscapes to better suit their needs and preferences. They may do this by planning and creatively placing different plant materials on their land. The choice of most Americans to do this has developed into the multibillion dollar industry of landscaping (Figure 8–2). This business employs thousands of people in many different areas of landscape development.

Landscape architects or **landscape designers** usually create landscapes. Both professions are

FIGURE 8–1

The term landscape actually means a view of any natural scenery as seen from one perspective.

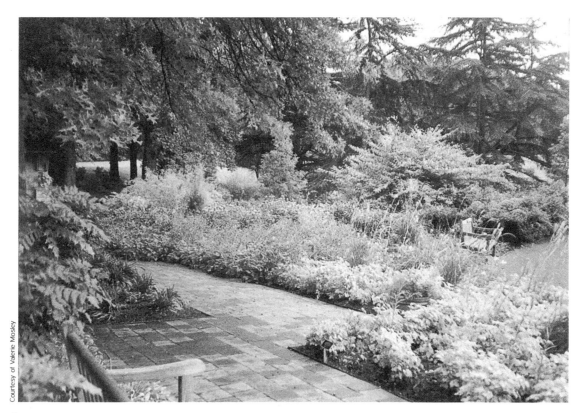

FIGURE 8-2

Landscaping has developed into a multibillion dollar industry.

trained in basic horticulture and design. They also work extensively with plants and other outdoor elements to create landscapes. However, the landscape architect differs slightly from the designer. The landscape architect is trained not only to design where roads, walkways, fountains, water drainage systems, and small outdoor structures will go but also in how they will be built. The most important skills of both professions are the ability to work well with other people, listen, and communicate their ideas through drawing, speaking, and writing (Figure 8-3).

Landscape contractors are the people who actually construct the landscape plans. They assemble and install all of the materials exactly as the landscape architect or designer has instructed in the plan. Landscape contractors must also be able to work well with other people and communicate, as they will often be in contact with the client or the landscape architect/designer. The contractor usually employs a team of landscape technicians to help install the design. Landscape contracting companies also have maintenance crews to help keep the landscape in

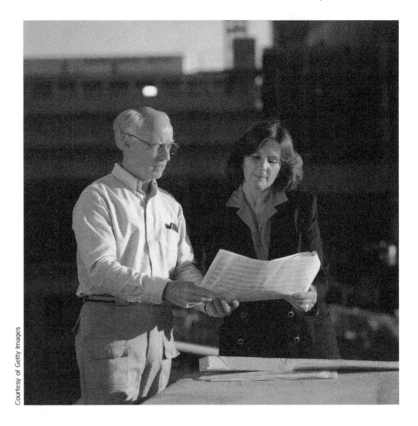

FIGURE 8-3
Landscape architects and designers must be able to work well with others, listen, and communicate their ideas through speaking, drawing, and writing.

good condition once installation is complete. Sometimes landscape architects and designers will form a landscape contracting company as well. In this way, the company not only designs the plans but also can install the new landscape.

Plant Materials

Landscape design is a complex art. Like many artists, the landscape architect or designer brings together various elements to create an attractive, functioning, and safe place. Several different materials are used in landscaping, such as rocks and water, and human-made features such as walkways, decks, and arbors are also included. However, in almost all schemes, plants are essential to landscape designs (Figure 8–4).

Plants are living organisms that have distinct shapes, habits, and needs. They also change as they grow. Many different types of plants may be used in landscape designs. Those most commonly used are called woody landscape plants and herbaceous ornamental plants.

FIGURE 8-4
Several different materials are used in landscaping such as rocks, water, and hardscape features. however, in almost all schemes, plants are essential to landscape designs.

Woody landscape plants are the foundations of most planting designs. The term "woody" indicates that the plant's stems and branches contain wood tissue. Most trees and shrubs are woody landscape plants. However, this category may be divided into two types of plants: deciduous and evergreen.

Deciduous trees and shrubs are woody landscape plants that completely shed their leaves each year during the fall or early winter. Before separating from the tree or shrub, the leaves begin a dying process in the fall. At this time, the leaf color may change from green to bright red, yellow, orange, or brown, depending on the type of tree. Some examples include oak and maple trees.

Evergreen trees and shrubs shed some of their leaves occasionally throughout the year; however, a great majority of the leaves remain on the plant. As the name says, evergreens are always green. This adds important color during the winter when most grasses have turned gray and other plants have lost their leaves. Although they do not offer the brilliant fall colors of deciduous plants, the evergreens can be effective year-round for screening unsightly views and breaking strong winds that blow through the site. Pine trees and holly shrubs are examples of evergreens.

Herbaceous ornamental plants are also frequently used in landscape designs. The stems and branches of herbaceous ornamentals are not woody. These plants usually die down at the end of the growing season, and therefore do not provide year-round interest in the landscape. However, during their growing season they may offer incredible blooms and color. These flowers not only add strong visual interest for a garden's visitors but also attract wildlife such as birds, butterflies, and bees. Three types of herbaceous ornamental plants are annuals, biennials, and perennials (Figure 8–5).

Annual plants live only for one growing season. During the spring and summer they will live and produce leaves and blooms. At the end of the growing season the plant will die and will not return.

FIGURE 8-5
Annual and perennial plants are sometimes used in designing a landscape.

Biennials are plants that will live for two years only, dying after the second growing season. Perennials will return for every growing season.

Phases of Landscape Development

There are four basic phases to landscape development:
1. Site inventory and analysis
2. Landscape design
3. Landscape construction
4. Landscape maintenance

Each step is important and should be conducted in the order shown.

Site Inventory and Analysis

The landscape architect or designer will first meet with the client, or project owner, to understand what they want and need in their project. A landscape project can be a variety of sizes. It can be a big project, such as a shopping center or a school, or it can be a smaller project, such as a backyard or a park. After this meeting, the designer begins the first step in landscape development: site inventory and analysis. This process involves visiting the site to identify what is currently there, such as any existing structures, trees, or other vegetation. It also should include understanding the wind directions, the slope (whether the site is flat or hilly), the soil type, and the way water flows over the project site.

Once the designer has inventoried what is there, the information will be analyzed. Certain questions should be asked at this stage, depending on the project. For example, what, if any, of the existing vegetation should be removed? What types of plants grow best in this kind of soil? Where is true south (plants placed on the south side of the site will receive the most sun)?

Every plot of ground is unique. Performing a site analysis helps the designer understand the character of the project site. This will ensure a more successful landscape design.

Landscape Design

When the designer has completed the site inventory and analysis, he or she moves to the landscape design phase. In this phase, the designer pulls together all of the information on the site with the needs of the client and forms a landscape design. The creation of the design takes a lot of time and goes through a long process of designing, evaluating, then redesigning and developing the plans.

The Design Principles There are five basic design principles that should be considered in designing landscapes: emphasis (focal point), balance, proportion and scale, rhythm, and variety and unity.

All of the principles are essential components. They must work together to create a strong, successful design. The ideas and artistic ability of the landscape architect or designer determine how each principle is used.

Emphasis is also called the focal point of the design. Every design should have some element that stands out enough to draw attention. Focal points communicate how people should move throughout the landscape (Figure 8–6). They also can show how

FIGURE 8-6
A focal point can lead visitors through the garden.

the landscape should be viewed. A focal point can be an unusual plant, a statue, a pool or fountain, or a spectacular view. Sometimes a focal point can be a building.

Although the point of emphasis should be strong enough to get attention, it should not be too strong. Otherwise, it overpowers the entire landscape design. Placing too many focal points will confuse people. The purpose of the focal point is to order the design by giving the eye a comfortable place to rest.

A balanced landscape is a comfortable landscape. This means that one part of the landscape design does not appear to weigh more than another area. There are two types of balance: symmetrical and asymmetrical. Symmetrical balance is the more formal type.

Both types of visual balance are used in landscape designs. If there is a really large, dark plant on one end of a building, the designer can locate another of the exact same plant in the same place on the opposite end of the building to balance it. This would be symmetrical balance (Figure 8–7). Using the same first plant, the designer could instead locate a group of about three smaller, lighter plants of a different type somewhere on the opposite end of the building. This contrast would create asymmetrical balance (Figure 8–8).

Proportion means that the plant material and human-made structures should generally be in scale

149
LANDSCAPING

FIGURE 8-7

This is an example of a symmetrically balanced residential landscape.

FIGURE 8-8

This landscape is asymmetrically balanced.

with each other. For example, a very large house should have large trees and shrubs nearest the house. It would look uncomfortable and out of scale if its landscape had only small trees and ground covers. Of course, people are different. Some may prefer using plants that are not in proportion to the building and other plants. For the most part, designing a well-proportioned landscape will create a more comfortable area (Figure 8–9).

Think of your favorite song. All songs have a particular rhythm that helps us remember them. A landscape works the same way. Just as musicians repeat chords in their songs, the landscape designer

FIGURE 8-9
These low-growing plants are in good proportion. They enhance the gentle slope of the ground.

can place the same type of plant in different places to establish a rhythm. For example, azaleas placed in certain places throughout the garden will give a sense of continuity. This brings us to the last principle of design: unity and variety.

Unity can be a very difficult principle to master, but it is extremely important. A brick or stone walkway, also called hardscape, can be unifying as it ambles from one area of the landscape to another. Establishing a rhythm through plants or hardscape materials will help to unify a design. All the elements of a landscape design should work together as a whole, and the different parts should not compete for attention (Figure 8–10). However, landscape architects or designers usually group variety with unity as the final design principle. This is because the two should check each other. When there is too much variety, unity is lost. When there is too much unity, the design becomes boring. Ideally, there should be just enough variety to keep the design appealing, yet enough unity so that the entire design works together.

The Landscape Plan After completing the design process, the landscape architect or designer will create a landscape plan to communicate his or her ideas to the client. A plan is any drawing that represents the site or subject area as if viewed from

FIGURE 8-10
A landscape design should work together as a whole. It should not have several very different parts that compete for attention.

FIGURE 8-11
A landscape plan is a drawing that represents the site or subject area as if viewed from above.

above (Figure 8-11). Imagine that you are hovering over your desk. If you were to draw what your desk looks like from that perspective, you would create a plan view of your desk. It is the same with a landscape plan. The landscape plan shows how the site has been designed from a plan view.

On this particular plan, the designer will identify walkways, roads, benches, water fountains, plant masses, or whatever has been included in the design. Landscape plans are usually colored, or rendered, to make them more understandable and attractive to the client.

The Planting Plan

Along with the landscape plan, the landscape architect or designer creates a planting plan. The landscape contractor uses this plan to lay out and install the landscape plan, so it must clearly communicate the designer's ideas. The planting plan is a simple drawing without color that focuses specifically on the plants in the design. This plan must indicate the placement, name, and number of every plant to be used in the design. On the same sheet of paper, the landscape architect or designer may include details, or small drawings, that show how they wish the trees and shrubs to be planted.

A planting schedule is also included in the planting plan. This is basically a small chart that adds up the plant information that has been labeled on the plan. A good, clearly organized planting schedule is very helpful to the landscape contractor. It will indicate all that is necessary for landscape construction. The schedule must include the total quantity, the botanical plant names, the common plant names, the size of the plants, and the spacing between plants at the time of planting. The designer also may include an area for any special notes concerning specific plants (Figure 8–12).

The botanical plant name is its Latin name. It is important that the designers know the botanical name to avoid confusion in locating the plant. Because common plant names have developed over time, based on common speech, the same plant may be called by different names in different areas. For example, the beautiful small tree with the botanical name *Chionanthus virginicus* is commonly called a fringe tree in the southern United States but a Grancy Grey Beard in the north. A sample planting schedule follows.

FIGURE 8-12
The planting schedule must include the total quantity, botanical plant names, common plant names, size of the plants, spacing between plants at the time of planting, and any important notes about a particular plant.

Quantity	Botanical Name	Common Name	Size	Spacing	Notes
20	*Chionanthus virginicus*	Fringe tree	15 gal.	20' on center	Multitrunk
18	*Itea virginica*	Virginia sweetspire	5 gal.	6' on center	
35	*Raphiolepsis indica*	India hawthorn	3 gal.	3' on center	

Landscape Construction

After the client has approved the landscape design and the plans have been completed, it is time to begin construction (Figure 8–13). Spring or fall is the best time to construct landscapes. Planting at these times gives the new plants time to grow and adjust to their new home before facing the drying heat of summer or the strong, cold winds of winter. If an irrigation system is to be added, it will be installed before any planting begins. The landscape contractor also will build any hardscape materials for the project, such as walls or walkways, before beginning to install the plants.

A copy of the landscape plan and the planting plan will be given to the landscape contractor whom the client has selected. The contractor will usually consult with the landscape architect or

FIGURE 8–13
After the plan is completed and approved, a contractor will construct the landscape. This worker is installing sod that will produce a nice grass lawn.

CAREER DEVELOPMENT EVENTS

The FFA Nursery/Landscape Career Development Event is a competition for teams consisting of four members. Each member competes individually and the final team score is the sum of the three highest individual scores. The contest consists of four parts: (1) identification of plant materials; (2) placing classes, problem solving, and equipment; (3) general knowledge; and (4) practicum.

Identification of plant materials consists of plants selected from the State Nursery/Landscape List and displayed for visual identification. Students may not touch the live plant material but can use a magnifying glass. The placing classes, problem solving, and equipment phase of the contest consists of five items with any combination of judging classes, problem solving, and questions about landscape/nursery equipment. The judging classes are plant materials or landscape plans in which the student places the materials from best to worst. Plant material is placed based on its structural form, market value, and health. The landscape plans should be placed by the criteria of design principles, plant selection and placement, and use of basic drafting methods. Nursery/landscape plants will be selected from the identification list and will be the same species and cultivar in each class. Plants are placed by structure, market value, and health.

The general knowledge part of the nursery/landscape event consists of 50 multiple-choice questions. This phase of the event tests for knowledge and understanding of all areas of the nursery/landscape industry. Areas for testing include plant materials, plant disorders, cultural practices, landscape design and construction, supplies and equipment, safety, equipment operation and maintenance, interpersonal relations, salesmanship, and records and reports. In the practicum, students identify plant disorders and answer questions about a given landscape plan. Plant disorders include diseases, insects, weeds, and nutritional disorders. Disorders are displayed for the student to view with the naked eye or a magnifying glass. The student is asked to identify the problem and those pesticides or other treatments necessary to remedy the problem. The landscape-plan practicum tests the student's ability to read a

LANDSCAPING

As part of the Nursery/Landscape Career Development Event, students must determine where plants are to be used in landscaping. *Courtesy of Carol Duval, National FFA Organization.*

landscape plan and knowledge of landscape design principles by asking multiple-choice questions.

After all team members have completed each phase of the event, each individual score is calculated. The top three individual scores are added to yield the total team score. If there is a tie between two team scores, the written exam score is used as a tiebreaker for first and second place. Winning teams from each state advance to the National Nursery/Landscape Event.

The Turf and Landscape Management Proficiency Award is an individual competition. This proficiency award can be applied for if an FFA member's job experience has focused on the installation, maintenance, and design of landscape plants and materials. Vincent is a senior and the winner of the chapter proficiency award for turf and landscape. He started his own lawn care service when he was in the eighth grade. He now has contracts with several local businesses and many home owners. Skills such as mowing, aerating, dethatching, fertilizing, and weed control are part of Vincent's business. Last year he made over $4,000 before his expenses were subtracted. Vincent was on the FFA senior high parliamentary procedure team and livestock show team, and served as chapter vice president. All proficiency awards are judged on the student's application and experience. Good record keeping of inventories and expenses plus the development of proficiency-related skills and accomplishments are essential to becoming a state and national winner.

designer if there are any questions, and will order all of the materials necessary to construct the landscape. Then, either the landscape architect, designer, or contractor will begin laying out the design according to the plan.

Once the design has been laid out on the project site, the contractor's team will begin preparing the soil for planting. When the soil is tilled and ready, installation begins. At the time of planting, some trees are so large that they require bulky equipment such as tractors or cranes to place them. It is a good idea to first locate and plant the largest materials, usually trees, to avoid damaging other plants. Next, the shrubs can be planted. The contractor finishes with smaller plants and the installation of sod or grass seed for the lawn.

Mulch is the key to healthy planting beds. After all planting is complete, the landscape team will mulch the planted areas with pine straw or wood chips at about 2 to 3 inches deep. This will insulate the plants during the winter months. It also will help keep the soil moist during the hot, dry summer months.

Landscape Maintenance

Once landscapes have been installed, they must be maintained to keep them healthy and attractive. Often a landscaping company or an individual offers landscape maintenance. This means that at certain times the company or person will mow the lawn, weed the beds, ensure irrigation systems are working properly, and check the plants for pest problems.

Landscape maintenance begins in the early spring and continues until late fall. It begins with a general cleanup followed by evaluation. This involves weeding and edging beds as necessary, removing dead leaves and branches from trees and shrubs, mowing the grass, and adding fresh mulch to the planting beds. The landscape maintenance person or technician uses a variety of tools during the cleanup. For small areas, simple hand tools and lawn mowers may be used. For the larger lawns of businesses or parks, bigger equipment is used such as motorized leaf blowers and tractors with a mowing blade attachment. A general evaluation is then done. The technician looks for pest damage on all of the plants, including the lawn.

Lawn Care Trees, shrubs, and herbaceous plants can require attention such as pruning, irrigation, and fertilization. But, probably the most time and money spent in landscape maintenance is on lawn care. Imagine large parks or residential landscapes that you have seen. Many of them likely have big, smooth, green areas of lawn. Although the lush lawns look simple, they don't grow and stay that way by themselves. Sometimes when a lawn is suffering from pest problems, large areas of the grass die out. A chemical treatment is usually needed if there is a serious pest problem (Figure 8–14). This must be done by licensed applicators. These people know the proper pesticide to use and the correct amounts. They must follow strict guidelines to prevent problems with the toxic materials.

Fertilizing is the next step. Grasses that grow in the cool part of the year receive their first fertilization in early spring. Those growing best during warmer weather are fertilized in late spring. Many people prefer that their lawns stay only one type of grass. When this is desired, a chemical weed killer is

FIGURE 8–14

Plants sometimes have to be treated with chemicals to control pests.

used to keep the other grasses and wildflowers out of the lawn. This weed killer is generally applied in combination with the first lawn fertilization.

Grasses also must get enough water to do their best. Water is usually added by irrigation. This can be done with overhead irrigation such as moveable sprinkler heads. Underground irrigation also can be used. This is when the water lines of an irrigation system are buried and only the sprinkler heads pop up when the water is turned on (Figure 8–15). Irrigation is especially important in the early growing season. If grass does not get the water it requires when it first starts growing, it may remain stunted for the rest of the year.

Mowing usually requires the most time spent on a well-maintained lawn. If the grass is healthy and lush, it will grow quickly. Many of you are probably familiar with a lawn mower. You may even mow the lawn weekly and can understand the amount of time mowing requires. To remain healthy, most grass needs to be mowed at 2½ to 4 inches in height (Figure 8–16). This depends on the type of grass and the height with which the home owner is comfortable. In most cases, such details are left for the maintenance person to decide.

With proper fertilization, pest control, irrigation, and mowing, a lawn can usually be kept beautiful for many years. Lawn areas can provide wonderful areas

FIGURE 8–15

Irrigation systems provide water for the landscape plants.

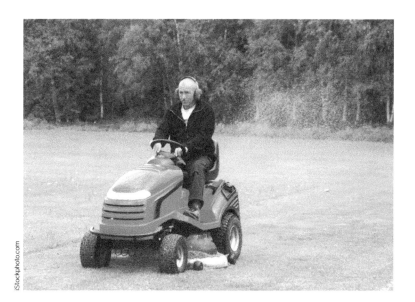

FIGURE 8–16
For grass to remain healthy and attractive, it has to be mowed at the proper height.

for playing, running, or relaxing outside. A lawn is much less expensive to install than areas planted with trees, shrubs, or ground covers. However, lawns also can be time consuming and expensive to maintain. For this reason, many home owners are now requesting landscape designs with smaller lawn areas and larger naturalized areas, mulched areas that are filled with plants that grow there naturally.

Pruning Pruning is the removal of selected branches of shrubs or trees. This is done to maintain the size or shape of a plant (Figure 8–17). Sometimes part of

FIGURE 8–17
Pruning is the removal of some of the branches. This helps to shape the tree or shrub.

the plant is damaged. This damaged portion must be removed. Afterward, the plant may need to be pruned into a more pleasing shape.

Pruning can be considered an art form. Many gardens in Europe have plants that are pruned into topiaries. Topiary can be anything from simple geometric forms to intricate animal shapes (Figure 8–18). Many painstaking hours go into these works of art. As the plants grow, they are pruned to allow certain parts to grow more than others. This gives the topiary its shape.

In landscape maintenance, pruning is one of the most important procedures to do correctly. A plant that is pruned badly might not recover, or it could actually die. Spring is the best time to prune plants; however, some also can be pruned in the winter.

Most plants that bloom in the spring should be pruned immediately after they have bloomed and not in the fall. Fall pruning cuts off buds that would normally bloom the following spring. Another problem with fall pruning is that pruning often stimulates a plant to produce new growth. This new growth is more susceptible to cold damage during the coming winter.

It is difficult to see the branching habit of deciduous plants in the summer. This makes it harder to decide correctly which branches to prune. The

FIGURE 8–18
Topiary can be anything from a simple geometric form to an intricate animal shape.

plants are also more susceptible to insect and disease damage and drought stress at this time of year. Therefore, most professional pruning is done during the late winter or early spring.

During the winter, pruners can see exactly what they are pruning. Also plants are in their dormant state in the winter months. This means that they are not actively growing above ground, though their roots and branches remain alive. Less damage is done to the plant when it is pruned in its dormant state.

Interiorscaping

Plants not only purify air and produce oxygen necessary to humans but also help create comfortable places for living, working, and shopping. As more and more people recognize the benefits of plants in human environments, landscape designs are being created for the indoors (Figure 8–19). Think about a mall or other place of business you have been to recently. Chances are there was some type of indoor landscape, or *interiorscape*, inside the area. An example of interiorscaping may include big potted plants, large bed areas, or gushing water fountains.

FIGURE 8–19

Interiorscapes bring the concepts of landscaping indoors.

The development process is the same for interiorscapes as for outdoor landscapes. However, there are some unique restrictions. The designer often does not have the freedom of expanding or moving planting beds. The height of the building also limits the size of the plants used.

Obviously, plants do not grow naturally inside. Light, soil, and water—everything necessary to a plant's health—must be provided. In the past, most buildings were not built with the health of plants in mind, and had to be modified to support plant life. Today, architects are beginning to design buildings that use natural light and provide areas for plantings. This makes it easier to include interiorscapes, although rarely is the lighting the same as that outside. In this instance, supplemental lighting must be provided. This means there must be a source of electricity as well as a place to anchor the lighting fixtures.

The choice of plants that can stay healthy inside buildings is limited. Most are tropical plants. These are plants that usually do well in extreme shade and can take warmer temperatures year-round. Fruit trees and most flowering shrubs cannot be grown inside because they must have a certain amount of cold before they will bloom.

Interior plants must be able to take a lot of abuse. Often the plants must tolerate indoor air pollution. Dust builds up on their leaves quickly in this environment. As a result, photosynthesis is reduced and the plants look bad. The plants must also have a reliable water source nearby. The architecture of the building or the plant containers must be designed to hold water seepage. Clients may get upset if they have water stains on their carpets. These factors must be considered as the plants are placed in each location.

Just like outdoor plants, indoor plants often need fertilizer. They do require less, however, and it should be applied less frequently. Fertilizing schedules have to be developed and implemented to optimize the beauty of the plants.

FIGURE 8-20
Interiorscapes require a lot of maintenance.

Insects and other pests can damage indoor plants. It is especially important to spot these problems before they become serious. Because of the closed environment in which they grow, pest problems can spread rapidly. Because of the high rate of foot traffic around the plants, the pesticides must be used carefully when a serious pest problem does occur.

When properly planned and maintained, indoor plantings can give a business, home, or office a healthier and more comfortable atmosphere (Figure 8-20).

SUMMARY

Landscaping is a complex and fascinating field that extends in many directions and requires diverse skills. Landscape architects, designers, contractors, technicians—all are crucial to the success of any landscape design, whether indoors or outdoors. Although hard work is necessary in all of these occupations, it is rewarding to see a beautiful job well done that will be enjoyed for many years.

CHAPTER REVIEW

✷ Student Learning Activities

1. Art, like landscapes, should contain the five elements of design. Go to the library and find a photograph, picture, or painting that you like. Try to identify the design elements in your picture. You may want to lay pieces of trace paper over the image and draw the different elements of your picture on the trace—that means you should have five different pieces of trace with one element sketched on each. When these sheets are put together, you should have a rough sketch of the original image. Explain your image and trace sheets to your classmates.

2. Locate what you consider to be the most beautiful landscape in your area. It can be a park, a home landscape, a natural area—whatever you like most.

Draw a sketch or take a photograph of the area. Why do you like it? Are the design elements present in the landscape? You may wish to use the trace overlay technique from exercise 1 to identify them. Share your findings and opinions with the class.

3. Closely examine the landscape around your home. If you do not have a front- or backyard, examine a park, your school's grounds, or a neighbor's yard. What do you like or dislike about it? Why? Can you identify any design principles?

4. Make a sketch of your school building and its surrounding grounds. How would you design the area? Think about the uses of the landscape space and who will be maintaining it. Create a landscape design for the area. In your sketch, identify the types of areas (play area, vegetable garden, wildlife areas, etc.) and the types of plants that you think should be in the landscape. You do not have to know the specific plant names, but you should label what kind of plants you want. For example you could label a tree as being small, deciduous, and flowering, with red fall color.

True/False

1. _____ The focal point of a design also can be called the point of emphasis.
2. _____ Perennial plants live for only one growing season and never return.
3. _____ The leaves of evergreen plants change color in the fall.
4. _____ The stems and branches of herbaceous ornamental plants contain wood tissue.
5. _____ Before beginning the design process, the landscape architect or designer should visit the project site to inventory and analyze its existing conditions.
6. _____ Mulching plants will insulate them from the cold and will protect them from drying out in the heat.
7. _____ Fall is the best time to prune plants.
8. _____ Fruit trees and flowering shrubs make excellent indoor plants.
9. _____ A planting plan is a simple drawing with labels, details of how the materials should be planted, and a planting schedule.
10. _____ Designers and contractors should always use the common names of plants to avoid confusion in finding them.

Multiple Choice

1. Before beginning the landscape design phase, it is very important to complete a
   a. landscape plan
   b. site review
   c. site analysis
   d. design element test

2. Identify the item that does not belong in the list below. A planting plan must include the
   a. Latin and common plant names
   b. number of each plant to be used
   c. size of the plants at time of planting
   d. name of the nursery that will provide the plants

3. A(n) _____ creates landscape designs including planting plans and hardscape plans, but also may design where and how roads, outdoor structures, fountains, and water drainage systems will be built.
   a. landscape designer
   b. landscape architect
   c. architect
   d. landscape gardener

4. Once the landscape design and plans are complete, they will be given to a _____, who will begin construction of the design.
   a. landscape contractor
   b. landscape architect
   c. landscape gardener
   d. landscape designer

5. The process of removing selected branches of trees and shrubs is called
   a. balancing
   b. pinching
   c. pruning
   d. tying

6. Planting designs that generally have a more formal, traditional appearance are
   a. asymetrically balanced
   b. symetrically balanced
   c. not balanced
   d. informally balanced

7. In the landscape maintenance business, the most time and money is spent annually on
    a. interiorscaping
    b. pruning
    c. irrigation
    d. lawn care

8. Most professional pruning occurs in the
    a. spring
    b. summer
    c. fall
    d. winter

9. Woody landscape plants that lose all of their leaves at one time during the fall are
    a. deciduous
    b. evergreen
    c. perennials
    d. annuals

10. To avoid damaging other plants when installing a landscape design, it is important to first plant
    a. shrubs
    b. ground covers
    c. sod or grass seed
    d. trees

✺ Discussion

1. Explain what is meant by landscaping.
2. What are the five basic landscaping design principles?
3. How is unity expressed?
4. What is a landscaping plan?
5. What is meant by landscape maintenance?

CHAPTER 9
Fruit and Nut Production

Student Objectives

When you have finished studying this chapter, you should be able to:

- Explain how consumers can buy fresh fruit all year long.
- List the major fruit and nut crops in the United States.
- Describe how pome fruits are produced.
- Describe how stone fruits are produced.
- Describe the citrus industry.
- Discuss nut production in the United States.

Key Terms

pome fruits
grafting
cultivars

stone fruits
citrus fruits

If you go into any large grocery store you can find a wide variety of fruit. Fresh fruit of many types is readily available. Apples, peaches, cherries, oranges, strawberries, pears, bananas, and many other fruits can be bought at any time of the year (Figure 9–1). This has not always been the case. Fresh fruit spoils very easily. Until relatively recently, people could only eat fresh fruit during the time between when the fruit ripened and when it spoiled. Fruit was preserved, but preserved fruit was never the same as fresh. The fruit was dried, canned, or made into jelly and preserves. Even though a lot of fruit is eaten fresh, much fruit is still preserved by these methods (Figure 9–2).

Modern transportation and storage techniques allow us to move fresh fruits long distances. In the

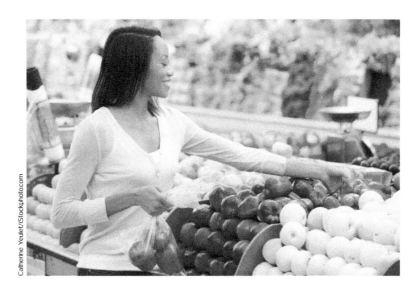

FIGURE 9-1

Many types of fresh fruit can be bought all year long.

FIGURE 9-2

Much fruit is still preserved.

FRUIT AND NUT PRODUCTION

Southern Hemisphere, the seasons are opposite from ours. When it is winter in North America, it is summer in that part of South America that is south of the equator. When our fruit trees are dormant, theirs are in full production. Fruit is shipped from South America and we can have fresh fruit in the middle of the winter. Peaches, pears, plums, and grapes are shipped to the United States during the winter. Most of the fresh apples we enjoy have been stored from the U.S. crop.

Pome Fruits

Pome fruits are those that have fleshy insides with small seeds. Only two pome fruits of importance are grown in this country—apples and pears. Most of the apples are grown in Washington, New York, Michigan, California, and the higher altitudes of the Appalachians. Apples grow so well in these areas because the trees need a cool climate. Cool nights during the ripening process make the apples have a good flavor and color (Figure 9–3). In the South, the nights are warmer during the time when the fruit ripens and the quality of the fruit is not as good.

About half of the apples produced are marketed fresh. The rest are processed into juice, jellies, pies, and other products. The most popular varieties are Red Delicious and Yellow Delicious, which are sold as fresh fruit. Other varieties include Macintosh, Rome Beauty, and Jonathan. These varieties are usually used for processing or cooking.

Most of the pears in this country are grown in Washington, Oregon, and California (Figure 9–4). Pears do well in the semiarid regions of these states. Pears are more difficult to produce in the East because of disease. A disease called fire blight attacks pear trees in the more humid areas. This blight is difficult to control and makes the pears expensive to produce. Many of the same practices used in the growing and managing of apples are used in growing pears.

FIGURE 9–3
Apples do best in climates where the nights are cool during ripening.

FIGURE 9–4
Most pears in this country are grown in Washington, Oregon, and California.

FIGURE 9-5
Pruning helps produce a more desirable tree.

Apple Production

The first step in the production of apples is the production of young trees. Instead of growing the trees from seed like many other plants, a process called grafting is used. Apples are difficult to propagate by any other method. For example, a hybrid apple tree might be difficult to grow from seed because the tree will be different from both of its parents. If a young sprout is combined with a portion of a desirable tree, the results can be a tree that grows and produces like the desirable tree (Figure 9-5). This method is also used to produce specialty trees such as dwarf trees and fruit trees that will grow several different varieties of fruit (Figure 9-6). Trees are also grafted onto rootstock that is stronger or more suited to the area.

Grafting involves cutting two parts of a tree. The lower part is called a rootstock and the upper part is called a scion (sometimes spelled "cion"). The rootstock may be larger than the scion. This gives the advantage of faster growth and earlier maturity. The two cuttings are carefully aligned and wrapped in string. The joint is then coated with wax and the plant is set in the soil. Eventually the two plant parts grow together and form a new tree. Through grafting, several varieties of apples can be grown on the same tree. Scions from different varieties are grafted on to different branches of a tree. Each grafted limb will grow the same type fruit as the tree from which the scion grew.

FIGURE 9-6
Through grafting, several varieties can be grown on the same tree. This process is done by cutting and joining woody tissue from different trees.

The young trees are set out in evenly spaced rows. Setting out in rows provides room for mowing and spraying machinery to maneuver. In arid areas, irrigation systems may be installed. The young trees are kept free of weeds and sprayed to prevent damage from insects and disease. Also, the young trees are pruned to shape them and to prevent the growth of weak branches.

It generally takes several years for the trees to begin bearing fruit. The length of time varies according to the variety, climate, soil conditions, management practices, and size of the trees. Because of the length of time involved before the trees produce, producers have a large investment in the trees (Figure 9–7).

Some trees, especially certain cultivars, are self-sterile. Cultivars are plants that are cultivated and grown by humans. These plants may have been bred up to the point where they are sterile unless pollinated from other plants. Red Delicious apple trees are a good example. By themselves, they are sterile and need another apple tree to cross-pollinate. Yellow Delicious trees produce pollen that is excellent in fertilizing Red Delicious trees.

Fruit trees are generally pollinated by insects (Figure 9–8). Remember from Chapter 4 that honeybees are particularly adept at pollinating.

FIGURE 9–7

An orchard represents a sizable investment before fruit is produced.

Fruit growers use bees to pollinate the trees. Hives are set in the orchards every year during the spring. The bees work the blossoms and also produce honey. Honey from fruit blossoms is delicious and is in great demand.

Once the trees have bloomed and the fruit has set, the trees have to be sprayed regularly to prevent insect and disease damage. Both disease and insect damage lower the quality of the fruit. Modern insecticides are safe when used correctly. The USDA carefully monitors the pesticide residues on fruit.

When the apples are ripe, harvesting begins. Most apples are harvested by hand to make sure the fruit is not bruised. Bruises lower both the appearance and flavor of the fruit. Pickers are careful how they handle the fruit (Figure 9–9). The fruit is

FIGURE 9-8
Most fruit trees have to be pollinated by bees or other insects.

FIGURE 9-9
Pickers have to be careful not to damage the fruit.

FRUIT AND NUT PRODUCTION

FIGURE 9-10
Properly stored, apples can be kept fresh for a year or more.

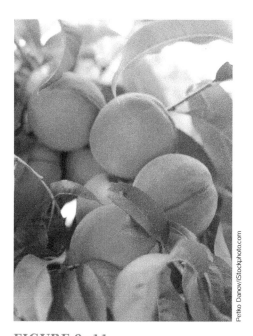

FIGURE 9-11
Peaches are grown all across the country. Different varieties grow in different areas.

gathered in small containers that do not damage the fruit.

Apples that are to be stored as fresh fruit receive special handling at the processing plant. Once the fruits are washed and cleaned, they are coated with a thin coating of paraffin to prevent damage to the skin and to help seal the surface of the fruit. The apples are stored in a process called controlled atmosphere storage. The fruit is placed in dark chambers where the temperature is held at a constant temperature of between 32 and 55 degrees F. Oxygen and carbon dioxide levels are decreased and are replaced by nitrogen. The apples are maintained in a proper balance of oxygen, carbon dioxide, and nitrogen. These gases, combined with the correct cool temperature and the absence of light, allow apples to be kept fresh for many months (Figure 9–10).

Stone Fruits

Stone fruits have large seeds called pits or stones in the center. The most commonly grown stone fruits are peaches, cherries, plums, and apricots. All of these fruits are grown worldwide and in the United States.

Peach Production

Peaches are grown all across the country (Figure 9–11). The major states are California, South Carolina, Georgia, New Jersey, Pennsylvania, and Michigan. Peaches need a certain number of hours of cold weather to bring the trees through the process of dormancy. Most varieties need around 750 hours of weather below 45 degrees F in order to have the proper amount of blossoms and fruit.

There are two types of peaches. One type is known as freestone. These varieties of peaches have stones that easily separate from the flesh when the peach is cut in half. The other type, clingstone, does not separate from the flesh of the peach. Most of the peaches that are marketed fresh are freestone and most clingstone are used for processing.

Like most fruit trees, peach trees are grafted onto rootstock. They are set out in rows and managed by pruning, fertilizing, and spraying. Peach trees are pruned differently than apples. Peach trees are pruned to have an open, inverted umbrella shape. This is to allow more sunlight into the branches of the tree (Figure 9–12).

Most peaches are harvested by hand. The fruit is so soft that it bruises very easily. Fruit that is intended for the fresh market must be free of bruises and blemishes. Peaches that are intended for processing may be harvested by a mechanical picker. These machines have a flexible platform that goes around the tree. The machine grabs the tree and gently shakes it. Ripe fruit falls onto the platform and the peaches are gathered in. The fruit is rushed to market where it is washed, peeled, and processed. The processing may include canning, pie production, or making into preserves.

FIGURE 9–12
Peach trees are pruned to have an open center.

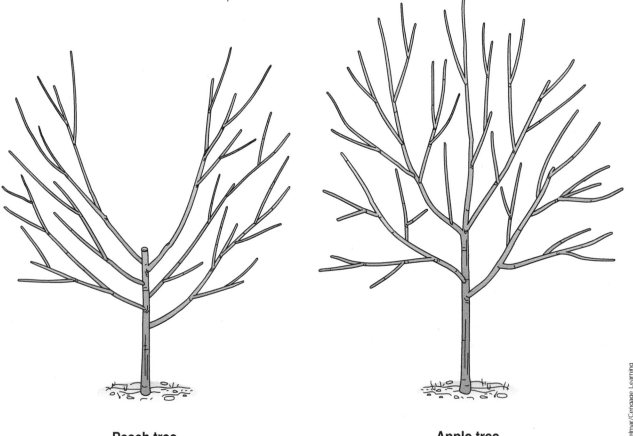

Peach tree **Apple tree**

Citrus Fruits

Citrus fruits have been grown for thousands of years. They are thought to have originated in Southeast Asia. They were introduced to Europe in the twelfth century and were brought to America by the European settlers. These fruits provide us with a good-tasting, nutritious addition to our diets. Citrus is an important source of vitamins and minerals, particularly vitamin C. During the 1750s, a British doctor named James Lind noticed that sailors who spent long tours at sea frequently came down with a disease called scurvy. The sailors were fed a diet of dried beans and dried meat. Dr. Lind discovered that by giving the sailors citrus fruit scurvy could be avoided. Later, it became known that the vitamin C in the fruit prevented the disease.

The citrus grown in the United States includes oranges, grapefruit, tangerines, tangelos, lemons, and limes. Most of the citrus is grown in Florida, California, Texas, and Arizona. These areas are about the only places in the United States with a climate warm enough to produce the fruit (Figure 9–13).

Although much of the crop is sold as fresh fruit, there is a large industry that processes

FIGURE 9–13

Citrus is grown in the extreme southern areas of the United States.

citrus. The largest processed products are orange, grapefruit, and lemon juices. The fruit is crushed to remove the juice. Most of the water is removed from the juice, and it is packaged and frozen. The consumer can then thaw the concentrate, add the proper amount of water, and enjoy orange juice (Figure 9–14).

SUPERVISED AGRICULTURAL EXPERIENCE

Everyone likes to receive recognition. The FFA Proficiency Awards and Supervised Agricultural Experience (SAE) are two ways every student can participate and become a winner. To begin, you must first become an FFA member and be enrolled in an agriculture class. Your teacher will help you design an SAE that will put into practice your interests plus the skills and abilities learned in the classroom, shop, and laboratory. By becoming active in an SAE, you will learn how to earn money that you may reinvest, spend, or save. You will also develop the skills required to produce, operate, and market your selected product. The Fruit and Vegetable Production Proficiency Award recognizes individual FFA members who develop and manage SAEs producing and marketing products such as field-grown vegetables, orchard fruit, and vineyard crops. Record keeping is a very important part of an SAE. The student must maintain a list of fruit and/or vegetable production skills that are learned in the classroom and/or used during the SAE. Records of development of equipment and material inventories plus the recording of income, expenses, and accomplishments must be accurate. The student must make managerial decisions involving the production and marketing of the fruit and/or vegetables and also provide the labor for planting, maintaining, and harvesting. An award-winning proficiency is the result of a long-term SAE, good record keeping, and a student's involvement with the local FFA and community.

Kerry won the Fruit and Vegetable Production Proficiency Award in his FFA chapter. He had grown a small vegetable garden in his first year in agriculture and won the chapter garden contest. In his second year he rented 1 acre of land and raised tomatoes and squash, and sold them at the local farmer's market. Kerry added the production of cool season crops (lettuce and cabbage) in his third year of agriculture. After the fourth year of his SAE, Kerry had made enough income to purchase 2 acres, where he plans to continue his vegetable business. Kerry was

One interesting by-product of the orange and grapefruit juice industry is the pulp. Pulp is the fibrous material left over when the juice is pressed out of the fruit. This material makes a valuable cattle feed. When mixed with grain, citrus pulp is used to fatten cattle for market. A good use has been found for a material that once went to waste.

on his high school football team and chaired the FFA Food for America program for two years. While attending the state FFA convention, Kerry placed second in the tractor-driving contest. Kerry's experiences with the FFA and his SAE have prepared him to be a successful leader, businessperson, and citizen.

The Fruit and Vegetable Production Proficiency Award recognizes individual FFA members who excel in producing and marketing fruits or vegetables. *Courtesy of Blane Marable, Morgan Co. High School, Madison, GA.*

FIGURE 9-14
Citrus juice is concentrated by removing most of the water. The juice is then frozen. Consumers mix water with the frozen concentrate.

 Nut Production

Nuts are actually seed kernels that are housed in a woody structure. They have been used as food for as long as people have been on the earth. Most nuts are high in proteins and fats. Nuts may contain as much as 600 to 700 calories per 100 grams, compared to 250 calories per 100 grams for beef with moderate fat content.

Each year producers in the United States harvest over 430,000 tons of nuts. The nuts are grown in the warmer regions of the country. While nut trees will grow in all states, they are generally sensitive to the cold and grow better in the South. Four types of nuts are of real commercial importance in the United States: almonds, walnuts, pecans, and filberts.

More almonds are produced in the United States than any other nut. Almost all of the almonds grown in this country are produced in California (Figure 9-15). The almond is native to the hot, arid regions of West Asia and California matches the climate. The almond tree resembles the peach tree. In fact, peach trees are often used as rootstock for grafting almond trees.

Although there are over 100 varieties of almonds grown, there are two major types of almonds. The sweet almond is edible and is grown for food. The

FIGURE 9-15
Almost all almonds grown in the United States are grown in California. This machine is shaking the nuts from the tree.

bitter almond is grown for the oil in the kernel. The sweet, edible almond is eaten raw, roasted, or in candies and pastries. The oil from the bitter almond is used to produce prussic acid, which is valuable as a fumigant. The oil is also used in some medicines.

The major type of walnut is the English walnut that is grown in California and Oregon (Figure 9–16). The black walnut is also grown, but it has smaller kernels that are more difficult to remove from the hull. Most English walnuts are grafted onto black walnut rootstock because the black walnut trees have stronger root systems.

Most walnuts are sold in the shell, although some are sold as kernels. These shelled nuts are used for baking and in salads. Another use for the walnut trees is the valuable wood. Walnut has always been a highly prized furniture wood.

Pecans are classified as genus *Carya,* which includes the hickories. Even though this nut tree is classified as a hickory, the nuts differ from other hickory nuts. Pecans have larger kernels and thinner shells. In fact, some pecan varieties are known as "paper shell pecans" because the very thin shells.

Pecans are grown in the South. The leading states are Georgia, Alabama, Texas, Louisiana, New Mexico, and Oklahoma. The trees grow quite large (75 to 90 feet) and mature slowly (Figure 9–17). Most need 8 to 10 years from the time they are set out

FIGURE 9–16

English walnuts are grown in California and Oregon.

FIGURE 9-17
Pecan trees grow quite large and mature slowly.

until they begin to produce. The trees live for many years and produce a long time.

Pecans are sold both shelled and unshelled. The nuts are used in pastries, pies, and other sweets. They are relatively high in fat content and contain a considerable amount of protein.

Filberts are also known as hazelnuts. Most are grown in Oregon and Washington. They are eaten raw or roasted and are becoming increasingly popular. One of the newer marketing techniques is to make consumers aware of the fact that filberts are hazelnuts. Consumers generally do not recognize the name filbert. The nuts grow on either a small tree or shrub, depending on the management practices used. The trees bloom in the winter, and the nuts mature in the fall.

Nuts are harvested by using a tree shaker. The machine grabs the tree and gently shakes it until the nuts fall off (Figure 9-18). Another machine sweeps the nuts into rows called windrows. A large vacuum picks the nuts off the ground (Figure 9-19). This means that the ground beneath the trees needs to be kept as clean as possible of weeds and foreign objects such as rocks. At the processing plant, the nuts are shaken to remove rocks, limbs, weeds, and any other foreign objects that were picked up. They are then placed in large bins, where the excess moisture is removed.

FRUIT AND NUT PRODUCTION

FIGURE 9-18
Nuts are removed from the tree by a machine that gently shakes the tree.

FIGURE 9-19
The nuts are picked up from the ground by a nut sweeper.

SUMMARY

The fruit and nut industry in the United States is very large. This industry provides consumers with food that is delicious and nutritious. Our lives are made better through the availability of fruits and nuts all year-round. In addition, they greatly add to our health.

CHAPTER REVIEW

Student Learning Activities

1. Visit a large grocery store and make a list of all the different types of fresh fruit that are available. Determine what country produced the fruits. Compare your findings with others in the class.

2. With your teacher's assistance, slice up several different varieties of apples and ask people in the class to taste the samples. Keep a list of the ones favored by the most people. Ask each person to tell why he or she preferred a particular variety.

3. Survey your area and determine what types of fruit and nuts are grown in your area. Report on the climatic conditions and other factors that make your area good for producing these crops.

4. Go to a large grocery store and find fruit you have never tried before. Purchase a sample and try it. If it is not a well-known fruit, try to explain why it is not more popular in your area.

True/False

1. _____ Fresh fruit spoils very easily.
2. _____ Most of the fresh apples we enjoy come from South America.
3. _____ Apple trees do best in cool climates.
4. _____ Fruit trees are generally pollinated by insects.
5. _____ Peaches, cherries, plums, and apricots are examples of stone fruits.
6. _____ Citrus pulp makes a good cattle feed.
7. _____ Nuts are grown in the cooler regions of the country.
8. _____ More pecans are produced in the United States than any other nut.
9. _____ Hazelnuts are also known as filberts.
10. _____ Nuts are harvested by using a tree shaker.

Multiple Choice

1. The only two pome fruits of importance that are grown in this country are apples and
    a. oranges
    b. pears
    c. cherries

2. The most popular variety of apple is
   a. Granny Smith
   b. Jonathan
   c. Red Delicious

3. Fruit growers make use of bees to pollinate the
   a. flowers
   b. seeds
   c. roots

4. The two types of peaches are freestone and
   a. filberts
   b. sandstone
   c. clingstone

5. Most of the peaches that are marketed fresh are
   a. freestone
   b. clingstone
   c. sandstone

6. Citrus fruits are thought to have originated in
   a. North America
   b. South Africa
   c. Southeast Asia

7. Most nuts are high in fats and
   a. filberts
   b. proteins
   c. seeds

8. Almost all of the almonds grown in the United States are produced in
   a. Washington
   b. Georgia
   c. California

9. The almond tree resembles the
   a. peach tree
   b. apple tree
   c. walnut tree

10. Pecans are grown in the
    a. West
    b. North
    c. South

Discussion

1. In the past, people could only eat fresh fruit during a very brief time of the year. Why?
2. What are some ways fruit is used?
3. Why are cool nights beneficial to apples?
4. What is fire blight?
5. What are the two types of peaches?
6. Why are peach trees pruned into an inverted umbrella shape?
7. How was the importance of vitamin C in citrus fruits discovered?
8. In the United States, where is most citrus fruit grown?
9. What three citrus fruits are processed most often?
10. What are the two types of almonds?

CHAPTER 10
Row Crops

Student Objectives

When you have finished studying this chapter, you should be able to:

- List the major row crops grown in the United States.
- List the types of grain that are produced in the United States.
- Describe the production and uses of grain crops.
- List the types of oilseed crops that are produced in the United States.
- Describe the production and uses of oil crops.
- List the types of fiber crops that are produced in the United States.
- Describe the production and uses of fiber crops.

Key Terms

row crops
grain
flour
cereal
oil crops
fiber crops

A large portion of our gigantic agricultural industry is row cropping. Row crops are those that are planted and cultivated in rows. Usually it refers to annual plants that have to be replanted each year. Every state in the country grows row crops. The type varies with the climate, soil type, and available markets. Examples of commonly grown row crops are soybeans, corn, cotton, peanuts, and wheat. Some crops, such as corn, are grown in all 50 states.

Hundreds of different types of crops are grown in this country. They may be classified as grains, oil crops, and fiber crops. It would be impossible to describe all of the different types of crops grown in this country. However, a few major crops comprise most of the row crops in this country. These major crops will be described in this chapter.

Grain Crops

Probably the very first crop that people planted was grain. Grain is the seed from plants that are classified as grasses. Humans discovered that not only were the grass seeds good to eat, but they could be planted and more grass that produced more seed would grow. The grasses that produced the most and best seed were selected for cultivation (Figure 10–1). As more was grown year after year, plants were developed that produced more seed that could be used for food.

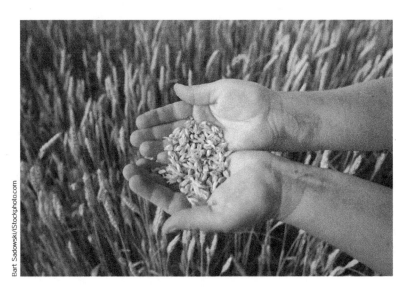

FIGURE 10–1

Grasses that produce the most and best seed are cultivated as grain. These seeds are from wheat plants before they mature enough for harvest

Wheat

Wheat is perhaps the oldest of the major grain crops. Ancient religious documents mention the growing of wheat. Most civilizations have grown wheat to be used in the making of bread. Bread is a staple food that can be preserved and provides nutrition.

Most of the wheat grown is used for human food. It has been the basis of bread for thousands of years. The seed kernels are ground into flour, and flour is used in baking. Wheat flour contains a high percentage of a substance called gluten, which causes the flour particles to stick together when moisture is added (Figure 10-2). This sticky dough traps gases from fermentation, which causes the bread to rise. When compared to other grains, wheat flour is relatively high in protein and makes a good-tasting food.

Wheat is grown in many countries of the world. It can be grown in a variety of climates, including cool climates. The United States produces about 12 percent of the world's wheat.

Wheat is grown all across the country, but most of the production is centered in the Midwest and Great Plains. Several types of wheat are grown in this country. These include hard red winter, soft red winter, hard red spring, white, and durum wheat.

Hard red winter wheat is grown in Kansas and nearby states. More hard red winter wheat is grown than any other type. It contains a relatively high

FIGURE 10-2

A substance in wheat flour called gluten makes it good for baking bread.

percentage of gluten that causes the flour to be among the best for baking breads. It is second only to hard red spring wheat for baking breads.

Soft red winter wheat is grown in areas of the Midwest where there is abundant rainfall. The moisture causes the wheat grains to be softer, more starchy, and lower in protein than the hard red winter wheat. Consequently, the gluten content is lower and the baking quality of the flour is not as good.

Hard red spring wheat has the highest amount of gluten of the commonly grown wheats. The highest quality breads are made from hard red spring wheat (Figure 10-3). This wheat is grown in the upper Great Plains area of North and South Dakota, Montana, and Minnesota. The weather gets too cold in these areas to grow winter wheat, so much of the wheat production is spring wheat.

White wheat is grown in the states of Washington and Oregon, with some being grown in Michigan and New York. White wheat is grown both as winter and spring wheat. The flour from white wheat does not make very good bread and is used mostly for making crackers, pastries, and breakfast cereal.

Durum wheat is grown through the upper Great Plains. It is more drought- and disease-resistant than the hard red winter wheat. Durum wheat contains a high percentage of gluten but it is considerably lower quality than the hard red winter and spring wheats. Durum is used for making pastas such as macaroni and spaghetti.

Winter wheat is planted in the fall and harvested in mid to late summer. Spring wheat is planted in the early spring and harvested in the late summer and early fall. At planting, the seeds are placed in the ground using an implement called a grain drill (Figure 10-4). The drill places seeds in close rows (6 to 12 inches apart) at the proper depth in the soil. Winter wheat must go through a period of prolonged cold before it can form grain heads. This process is called vernalization.

FIGURE 10-3

Hard red spring wheat makes the best flour for bread.

FIGURE 10-4

Wheat is planted using an implement called a grain drill.

FIGURE 10-5

Grain is harvested using a combine. The machine "combines" the process of cutting and threshing the grain.

Wheat harvesting is done with a combine (Figure 10-5). This machine cuts the wheat stalks off and threshes the grains of wheat out of the seed heads. A large fan blower separates the hulls and chaff from the grain. The grain is then conveyed to a bin that when full is emptied into a truck. The truck takes the grain to an elevator where it is sampled, graded, and stored. The producer is paid on the basis of the grade of the wheat. The grade is based on the moisture content and the amount of weed seed and other foreign matter in the grain.

Corn

More corn is produced in the United States than any other crop. The United States produces almost half of all the corn produced in the world. It is one of the few major crops that is native to this country. Before the European settlers arrived in the New World, corn was not grown in other parts of the world. However, it was widely cultivated by Native Americans in both North and South America.

Corn is grown in every state in the country, but the majority of the production is in the Midwest (Figure 10–6). In this region, the soil is deep and rich and the rainfall is adequate. This combination allows for huge yields of corn. These yields have dramatically increased over the past few decades. Research into methods of production has greatly increased production. However, the greatest impact has come from hybrid varieties. A hybrid is a cross between two different varieties. The resulting plant outproduces either of the parents. In 1930, when very little hybrid corn was grown, the average production was around 25 bushels per acre. Today, when almost all corn is hybrid varieties, the average yield is almost 150 bushels per acre. This average is for all of the country. A well-managed field in the Midwest may average well over 200 bushels per acre (Figure 10–7).

FIGURE 10–6

Corn is grown in every state in the United States.

FIGURE 10-7

A well-managed field of corn in the Midwest may average as much as 200 bushels of corn per acre.

Corn is used both for human food and animal feed. Almost all of the corn produced is either white or yellow. Of the two types, yellow corn is by far the most widely grown. White corn is grown for corn meal that is used in baking and bread making. It is also used in making breakfast cereals and corn chips because it toasts better than yellow corn. Yellow corn contains substances called carotenoid pigments that animals can convert to vitamin A. Both yellow and white corn are used for animal feed because they are so high in carbohydrates (Figure 10-8). Cattle and hogs are fattened on corn because it is so high in this nutrient. Because corn is relatively low in protein, a protein supplement is usually added to livestock feed.

Corn has other uses. Cooking oil is pressed from the grain. It has no cholesterol and is of very high quality. Cornstarch is produced from the starchy content of the grain. It is used in thickening agents in puddings and pies, as a drying agent in body powder, and for a broad array of other uses.

A major new use of corn is the production of ethyl alcohol, or ethanol. Alcohol made from corn and other grains can be made to run car, truck, tractor, and other engines as efficiently as gasoline

FIGURE 10-8
White corn is used more for human food. Yellow corn is used more for animal feed.

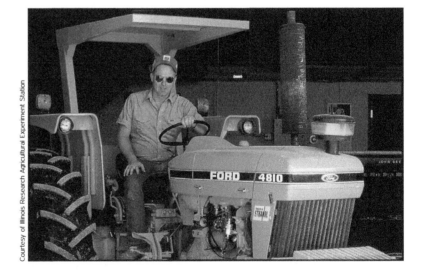

FIGURE 10-9
Some engines make use of ethanol that is made from corn.

(Figure 10-9). The technology for this process is hundreds of years old and is really a simple procedure. Alcohol produced from corn holds promise for replacing much of the petroleum-based fuels used today. Since petroleum products are considered to be nonrenewable resources (resources that cannot be replaced), ways of producing crops that can be converted to fuel have been explored. At our high rate of usage, petroleum oil will someday run out. A mixture of ethanol and gasoline called gasohol now replaces up to 10 percent of the gasoline used in the United States. More about this use will be discussed in Chapter 23.

Oil Crops

A major part of the row crops in this country is produced for their oil content. The two most widely produced oil crops are soybeans and peanuts. The oil from the seeds of these plants is used in a variety of ways ranging from cooking oils to paints to a replacement for diesel fuel (Figure 10–10).

Soybeans

Soybeans probably originated in China, where they have been grown for thousands of years. They were imported into this country in the late 1800s. Through the development of varieties suited to the climate of various parts of the United States, soybeans have become a major crop in this country. In fact, the United States produces over half of the world's supply of soybeans (Figure 10–11). It is interesting that the United States now exports soybeans to China,

FIGURE 10–10
Oil from the seed of plants is used for cooking oil. This is sunflower seed that is pressed to obtain oil for cooking.

FIGURE 10–11
The United States produces over half the world's supply of soybeans.

the country in which the crop originated. This crop is grown extensively throughout the South and Midwest.

One reason that soybeans are popular is the wide variety of uses. One of the primary uses is for

SUPERVISED AGRICULTURAL EXPERIENCE

Proficiency Awards

FFA members have a great opportunity to compete for awards in the area of row crops, including grain, oil, and fiber products. These awards encourage the use of new technology and management practices in the production and marketing of agricultural crops. By taking agriculture classes, the student can learn and practice the knowledge and skills required to become an effective producer of agricultural products. Producing crops is a great Supervised Agricultural Experience (SAE) for any FFA member. The agriculture teacher, student, and the student's guardian should agree on the scope and focus of any SAE. Students will be responsible for making managerial decisions about the ground preparation, planting, maintenance, harvesting, and marketing of the crop. The student must keep a list of crop production skills used in the classroom and in the field. Income, expenses, and yields of the crops produced and harvested also must be accurately recorded. The building of inventory such as land, seed, fertilizer, machinery, and harvested crops over several years is essential to an award-winning proficiency.

Jill is an active member of her FFA. In her senior year she won the state agriculture management contest and a scholarship to her state's agriculture college. Jill began her SAE in the ninth grade. She lived on a 100-acre family farm, and part of the signed agreement between herself, her father, and her agriculture teacher was to lease 2 acres of land to Jill. During the first two years Jill tilled, sowed, managed, and harvested cotton on the 2 acres. Jill's agribusiness class taught her record-keeping skills and investment opportunities that helped her reinvest her income wisely. Jill chose to sell her cotton with a local co-op. The last two years Jill diversified her production by crop rotations. After paying all the expenses of her operations, Jill's profit margin was small, but the experience was priceless. Jill's application for the crop proficiency award won chapter, district, and state recognition. Jill is one of many who have found the FFA to be the foundation of her future.

Frank is a high school student who lives in the city. This year he is applying for the chapter placement in agricultural production proficiency. Frank has been employed on oil. The oil is pressed out and used for cooking and making margarine, shortening, cheese spreads, and coffee whiteners. New uses for the oil, such as a substitute for diesel fuel and as the basis for

a forage production farm for the last three years. Frank's FFA membership started in the seventh grade. He continued to be an active member and competed on the forestry and land-judging teams. In his sophomore year Frank came to his teacher looking for an after-school job. Frank's desire to be outdoors and his accomplishments in agriculture mechanics class led his teacher to place him on a local forage-production farm just outside the city. Frank has kept accurate records of his hours of work, job skills, income, and expenses related to his job. Through the years, Frank was promoted from field hand to large-equipment operator. His experiences and skills have prepared Frank for a career in agriculture, and he is an excellent candidate for the proficiency placement award in agricultural production.

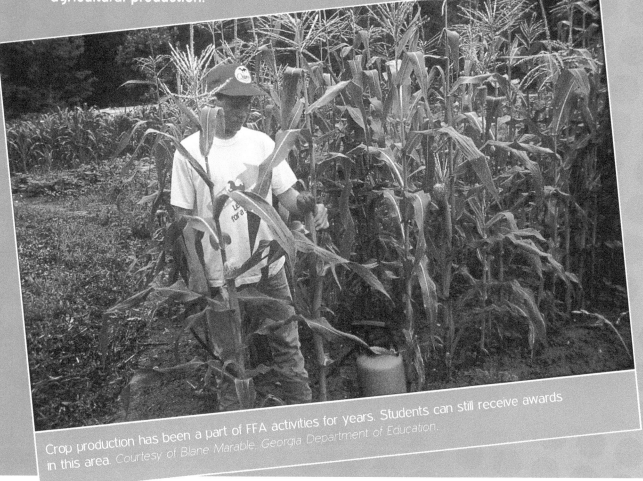

Crop production has been a part of FFA activities for years. Students can still receive awards in this area. *Courtesy of Blane Marable, Georgia Department of Education.*

a high-quality ink, have been discovered (Figure 10-12). The whole bean can be used to make meat substitutes and extenders.

The material left when the oil is pressed out is used for animal feed. The pressed cake is ground into a meal that contains an average of 45 percent protein. This high-quality feed is mixed with corn and other grains and fed to pigs and chickens.

Peanuts

The peanut is native to South America and was transported to Africa by Europeans. From there, it was brought to North America by Africans who were brought in as slaves. This nutritious food gained popularity in the latter part of the 1800s, when the boll weevil destroyed much of the cotton crop of the South. Scientists such as George Washington Carver invented new uses for the peanut. Products such as peanut butter became popular and a new industry grew up in the South.

Today, this crop is grown in many parts of the South. The leading states are Georgia, Alabama, Texas, and North Carolina. The nuts grow on "pegs" that extend from the stem into the ground (Figure 10-13). They are planted in rows about

FIGURE 10-12
Soybeans are used for a variety of purposes ranging from animal feed to printer's ink.

FIGURE 10-13
Peanuts grow from "pegs" that extend from the nodes of the stem into the ground. The pods grow underground.

FIGURE 10-14

A large plow turns the peanuts toward the sky.

3 feet apart. At the time of harvest, the vines may completely cover the ground. The plants require a long growing season with warm weather. In addition, because the nuts grow in the ground, they must be grown in light sandy soil. Peanuts grown in soils with heavy clays are difficult to harvest.

At harvest, a large plow turns the roots toward the sky (Figure 10-14). The vines are left in the field for several days until the nuts begin to dry. A large machine picks up the vines and removes the nuts. The nuts are allowed to dry and are stored or processed.

Peanuts are sold in the shell raw, shelled raw, roasted in the shell, shelled and roasted, or boiled. Products such as peanut brittle, a variety of candies, and pastries are made from peanuts. Of course, many of us enjoy eating peanut butter, which is made from ground-up roasted peanuts.

One of the most valuable products from peanuts is peanut oil. This is one of the highest-quality cooking oils. Like soybeans, peanuts are pressed by machines until the oil is removed. The cake that is left is used as a high-quality livestock feed. Not only is it high in protein content (around 50 percent), but also the protein is very high quality. In addition, the dried plants are often fed to cattle as hay.

Fiber Crops

Much of the cloth produced today (such as rayon, nylon, and polyester) is manufactured from petroleum. In a process that was developed in the first half of the twentieth century, petroleum is processed into long fibers that are used to make cloth. This process gives us cloth that is relatively inexpensive and very durable. For these reasons, fiber crops declined in popularity for several years. However, today, natural fibers are rebounding as a clothing material. Human-made fibers may be more wrinkle resistant and durable but for comfort they cannot match fiber produced by agriculture (Figure 10–15). Also, a lot of cloth produced is a combination (called a blend) of artificial and natural fibers that gives the flexibility and durability of synthetics and the comfort of natural fibers. The use of natural fibers makes us less dependent on imported petroleum.

Two main kinds of natural plant fibers produced for cloth are grown in the United States. These are cotton and flax, and both of these are also used as oil crops.

Cotton

Cotton was probably first used in the Nile Valley in ancient Egypt. The climate and the fertile soil were ideal for the growth of the plant. In North America, humans have

FIGURE 10–15
Human-made fibers cannot match cotton clothing for comfort.

used cotton for at least 7,000 years, and its cultivation goes back hundreds of years before the European settlers arrived. Much of the history of the United States developed around the growing of cotton.

Cotton is a shrub like perennial plant (although it is cultivated as an annual plant) that requires a long growing season and warm temperatures (Figure 10–16). This means that the crop must be grown in the southern parts of this country. The leading states are California, Texas, Arizona, and the states of the lower Southeast.

About midseason, the cotton flower blooms from a bud that appears on the stem. As it opens, the flower is a pale yellow that gradually turns pink and then red (Figure 10–17). After the flower is fertilized, the seedpod, known as a boll, forms. This structure is shaped like a small football and contains the fiber and the seed. The seed grows within the boll and the fibers form on the seeds.

Cotton is picked by large machines that use rotating fingerlike projections on opposite rotating drums to remove the lint from the bolls

FIGURE 10–16
Cotton requires a long growing season.

FIGURE 10-17
Cotton blossoms are first yellow, then turn pink.

FIGURE 10-18
Cotton is picked by large machines.

(Figure 10–18). This ingenious invention replaced long, strenuous hours of manual labor that was once required to handpick the cotton. The lint is removed from the picker and is either stored in the field in long loaf-shaped modules that hold 10 to 12 bales of cotton, or it is placed in trailers and taken directly to the gin. A bale weighs around 1,200 pounds before the seeds are removed and around 500 pounds after the seeds are removed.

At the gin, the seeds are removed and the cotton is cleaned of trash and foreign material. Once the seeds are separated and the cotton is cleaned, the lint is pressed into 500-pound bales, graded, and stored or sent to the mill. The seeds are a secondary industry. They are sent to a processor where they are pressed to remove the oil. The oil is used as a cooking oil or in a variety of products from margarine to salad dressing. The cake that is left over after the oil is pressed out is ground into a meal that is a valuable source of protein. Cottonseed meal is fed to cattle and other ruminants. It cannot be utilized by pigs and certain other simple-stomach animals because of a toxic substance in the seeds called gossypol.

Flax

The use of linen dates back to the Stone Age when humans still wore skins for clothing but used linen for making fishnets. Over the years, linen has been used for clothing in most countries of the world. Much of its popularity as a fabric for clothes was lost with the widespread availability of cotton that began in the early 1800s. Today it is still popular as a cloth for making tablecloths, napkins, and in recent years has regained some popularity as a clothing material.

Linen comes from fibers produced in a plant called flax that grows in climates with plenty of rain and no high temperatures (Figure 10–19).

FIGURE 10–19

A high-quality cloth called linen is made from the flax plant.

Most of today's flax is produced in Europe and New Zealand. However, substantial amounts are produced in North and South Dakota and Minnesota.

Plant stems contain two structures, the xylem and the phloem, that transport water and nutrients through the plant. Linen comes from the fibers (called bast fibers) that make up the phloem of the plant.

The plants grow to a height of about 3 feet and are harvested. The outer layer of woody material must be dissolved before the fibers can be removed. This is done by soaking the stems in warm water where bacterial action decays the material or in a more modern method where the covering is dissolved using chemicals. This process is called retting.

After the outer layer is removed, the phloem fibers are passed through fluted rollers that break up the woody substances connected to the fibers. In a process called scutching, the usable fibers are separated out. Then the fibers are combed out in much the same process used in carding cotton. This lays out the fibers parallel to each other so they can be spun into yarn.

The seeds of the flax plant are pressed for their oil. This valuable oil, called linseed oil, is used in the manufacture of paints and varnishes. As with the other oilseeds, the cake left from oil removal is used for livestock feed.

SUMMARY

The production of row crops in the United States is a gigantic business. No other country in the world can compare to the United States in terms of the amount or variety of crops that are produced. Much of the wealth of the nation is a result of our crop production. In the future, perhaps, new crops and new uses for old crops will be developed.

10 CHAPTER REVIEW

Student Learning Activities

1. Examine the labels on foods in your home. List all the foods that contain grain or grain products. Remember that gluten and starch come from grain. Compare your list with others in the class.

2. Visit a grocery store and list all of the products that are made from wheat. Try to determine which type of wheat would be best suited for each product.

3. List all of the products in your home that were made from oil from seeds. Try to think of other ways oil from plants might be used someday. Share your ideas with the class.

4. Determine the top row crops grown in your state. List reasons why these particular crops are the ones most widely grown.

True/False

1. _____ Probably the very first crops that people planted were grains.
2. _____ Most of the wheat grown is used for human food.
3. _____ Most wheat is produced in the Northeast.
4. _____ More corn is produced in the United States than any other crop.
5. _____ The majority of corn production is in the Midwest.
6. _____ Much of the cloth produced today is manufactured from petroleum.
7. _____ Human-made fibers are more comfortable than fibers produced by agriculture.
8. _____ Cotton was probably first used in the Nile Valley in ancient Egypt.
9. _____ One of the most valuable products from peanuts is peanut oil.
10. _____ China produces over half of the world's supply of soybeans.

Multiple Choice

1. A large portion of our agriculture industry is
   a. row cropping
   b. corn production
   c. fruit production

2. The grasses that produce the most and best seeds are selected for
   a. grazing
   b. cultivation
   c. flax

3. The oldest of the major grain groups is perhaps
   a. wheat
   b. rice
   c. corn

4. Wheat flour is relatively high in
   a. fat
   b. sugar
   c. protein

5. Gasohol is a mixture of gasoline and
   a. ethanol
   b. diesel fuel
   c. propane

6. The two most widely produced oil crops are soybeans and
   a. corn
   b. olives
   c. peanuts

7. A shrublike plant that requires a long growing season and warm temperatures is
   a. flax
   b. cotton
   c. linen

8. Linen comes from fibers produced in a plant called
   a. gluten
   b. cotton
   c. flax

9. Plant stems contain these two structures:
   a. xylem and phloem
   b. xylem and gluten
   c. phloem and gluten

10. The seed of the flax plant is pressed for
    a. gluten
    b. linen
    c. oil

Discussion

1. What factors determine what row crop is grown in each state?
2. What are the three classifications of crops?
3. What are the types of wheat?
4. Why is hard red winter wheat so popular?
5. Durum wheat is used for what purpose?
6. Other than color, what is the difference between white and yellow corn?
7. What are some uses for corn?
8. What functions are served by the xylem and phloem?
9. Where is the majority of corn grown? Why?
10. What uses are made of soybean oil?

CHAPTER 11
Forest Science

Student Objectives

When you have finished studying this chapter, you should be able to:
- Discuss the extent of the forest industry.
- Explain why forestry can be considered agriculture.
- Explain the concept of natural succession.
- Discuss the forest as an ecosystem.
- Discuss the importance of the forests to humans.
- Discuss the production steps in growing trees.

Key Terms

forestry
wildlife habitat
hardwood
softwood
succession
canopy
wilderness areas
tree farms

FIGURE 11-1
Like any crop, trees are planted, managed, and harvested. These seedling pine trees are planted using a tractor and plow.

FIGURE 11-2
Each year, about 27 percent more trees are planted than are harvested. This forester is planting seedlings after harvesting trees.

One of the largest industries in agriculture is forestry. At one time forests were considered natural resources. Giant forests once stretched across the country and people made use of the trees for building homes. When it became apparent that these forests would not last forever, the cultivating of trees began. Trees are now grown very much the same as crops are grown. They are planted, managed, and harvested (Figure 11-1). Once harvest is complete, new trees are planted and a new crop is grown.

The forestry industry stretches from coast to coast and encompasses almost 500 million acres of trees. From these trees come thousands of different types of products that are manufactured using wood fiber. Over a million and a half people are employed in the management, harvesting, processing, and marketing of these products.

Some people think that the harvesting of timber is a horrible practice that destroys nature. This is an erroneous concept. Because of the conservation efforts of the forestry industry and the U.S. Forestry Service, today there are more trees in this country than there were 100 years ago. Each year close to two billion trees are planted in the United States. This represents more than six new trees a year for every American. Through the replanting of the forests, new plant growth flourishes. Each year about 27 percent more trees are planted than are harvested (Figure 11-2). Areas that once were used for row cropping are now used for growing trees. Hilly land not suited for farming with modern machinery is now planted in pine trees and is once again productive land.

Wildlife greatly benefits from the forest industry. As new trees are planted, lush growths of plants grow with the seedling trees. These plants provide food and habitat for wildlife. Several species of animals have been reintroduced where they had ceased to exist (Figure 11-3). For example, the white-tailed deer has made a tremendous comeback in recent

FIGURE 11-3
Today, there are more of some types of wildlife in the forests than ever before.

FIGURE 11-4
Some object to the cutting of large, old-growth trees. This huge tree is over 400 years old.

years. Some sources say that there are now more deer in this country than before the first European settlers arrived. A few decades ago, the wild turkey was on the verge of extinction because of the loss of its habitat and overhunting. Today, there are over four million wild turkeys living and thriving in our forests.

Most of the trees that are harvested in the United States were planted or have grown up from forests that were cut. Only in the Pacific Northwest are there large tracts of old-growth timber. This timber is also call virgin timber and has never been cut. Even today, timber companies are cutting old-growth timber in the Pacific Northwest. No finer construction timber exists. Homes all over the world are made from lumber cut from the trees in the Pacific Northwest. Many people object to cutting the huge old trees (Figure 11-4). A constant struggle seems to be ongoing between preservationists and the logging industry. The preservationists insist that as much of the old-growth timber as possible must be preserved. The logging industry points out that there are over four million acres of old-growth timber preserved

FIGURE 11-5
Logging operations were stopped in efforts to save the spotted owl.

FIGURE 11-6
A hardwood is a deciduous tree. This is an American Beech tree.

in this country that, by law, can never be cut. They contend that the country's housing industry needs the material from the large trees that are in the areas where harvesting is permitted. They further argue that if logging is stopped, many jobs will be lost as a result.

Another point of controversy is over **wildlife habitat**. A good example is the spotted owl that lives in the Pacific Northwest. Some biologists contended that the owl needed large tracts of old-growth timber in which to breed and raise their young. Logging operations were halted and many people lost their jobs in the effort to preserve the owl (Figure 11-5). The issue was debated over whether preserving the owl was more important than providing jobs. Later evidence showed that the owls could reproduce in second-growth forests.

The Natural Forest

At one time, trees cut for lumber came from the natural forest. Different types of forest grew in different regions of the country. Basically, there are two classifications of timber: **hardwood** and **softwood**. A hardwood is a tree that is a broadleaf deciduous (sheds its leaves in the winter) tree (Figure 11-6). A softwood is a cone-bearing, evergreen tree called a conifer (Figure 11-7). Actually, the classification has little to do with the hardness or softness of the woods. Some hardwoods, such as yellow poplar, are softer than softwoods like southern pine.

The upper regions of the Northeast at one time were covered in vast hardwood forests. Throughout the Midwest and the upper regions of the South, hardwoods also grew. Along the coastal plains of the South, the predominate trees were pines. The Rocky Mountain region was covered with conifers such as lodgepole and ponderosa pine. On the West Coast, gigantic softwoods such as Douglas fir, spruce, and redwood trees covered the slopes.

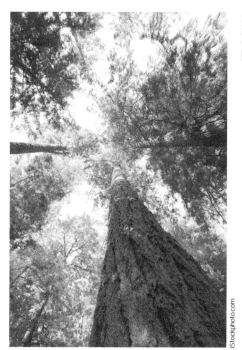

FIGURE 11-7
A softwood is an evergreen, cone-bearing tree. These are Douglas fir trees.

 Forest Succession

The reason certain types or species of trees dominate a region is because of a natural process called **succession**. When a forest first begins from bare ground as a result of fire, clear-cutting, or other causes, a lot of different plants germinate. These plants all compete for water, nutrients, and sunlight. The trees grow taller than other plants so eventually trees shade the other plants. Plants that do not receive enough sunlight generally stop growing and die. Trees that grow well in a particular climate are usually the ones that grow faster than the other species of trees. Also, some species just naturally grow faster than other species. The limbs of the faster-growing trees eventually form what is called a **canopy**, or overstory, above the other trees (Figure 11-8). This prevents trees of the same species from growing in competition. Trees that are not tolerant of shade will not grow under the canopy. However, trees that are tolerant of shade continue to grow at a slow pace.

Eventually the trees that are shade tolerant may overtake and suppress the trees that first created the overstory. This is particularly true if the first trees mature and die. A good example of this process is the forests that covered the upper South. The first trees to dominate were the fast-growing pines such as the

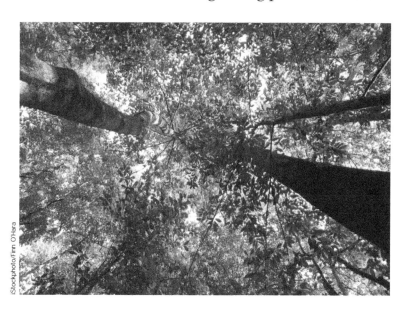

FIGURE 11-8
The limbs of faster-growing trees unite to form a canopy.

FORESTRY FIELD EVENTS

Many states hold forestry field days. The forestry field day is both an individual and a team competition. Team members train for skills used in today's forestry industry (see photo). Tree planting, forest management, tree disorders, timber-stand improvement, and estimating pulpwood and saw timber volumes are a few of the skills. Many forestry management practices are included in the FFA Forestry Event. Contestants must understand and answer forestry management questions concerning stand classification, merchantable trees, basal area, and cutting practices. Team members must practice and be ready to give estimations of pulpwood and saw timber standing in a given area, by measuring the tree heights and diameters. Testing the contestant's ability to improve a timber stand by harvesting select trees because of insect infestation, disease, lack of vigor, or tree conformation is another part of the event.

Foresters must develop good math skills. Many measurements and calculations are required to manage timberland efficiently and effectively. The event tests the student's ability to make ocular estimates of trees and calculate the number of acres and placement of boundary lines in a given area. The FFA team must read a compass and decide points and distances during the compass course. Members must identify not only trees but also the insects and diseases that attack the trees. The FFA Forestry Event requires a winning team to spend time studying classroom materials and practicing skills. Winning teams from each state compete at the National Forestry Contest.

The Forest Management Proficiency Awards encourage the use of forestry practices that conserve and increase the productivity and economic value of land. The award is presented to individual FFA members who have excelled in experiences using the principles and practices of today's forestry industry. To apply, the student must be involved in a forestry-related Supervised Agricultural Experience (SAE). The student must maintain a list of forest management skills learned and practiced during the project. The student also must develop an inventory of buildings, equipment, tools, and materials used in the experience program and keep

FOREST SCIENCE

Forestry events allow students to display their skills in forestry measurement and management. *Courtesy of Blane Marable, Georgia Agricultural Education.*

accurate records of income and expenses of forestry products produced, harvested, and sold. By keeping a list of accomplishments such as land preparation, planting, seeding, thinning, and harvesting, the student can show how he or she has improved the land use for forestry products.

Maria won the Forestry Management Proficiency Award in her FFA chapter. Her program in forestry developed from her interest in her agriculture forestry class and part-time job at a tree nursery. Over a four-year period, Maria recorded 856 hours working for the nursery and developed skills in customer service, propagation, planting, and maintenance of trees. Her experience at the nursery gave her the idea to propagate and grow Leyland cypress as Christmas trees. At her home she rented an acre of land from her parents and planted 200 one-gallon Leyland cypress trees that she bought from her employer. The first year, she also built a 10' x 15' cold frame and began propagating cuttings. During the second and third years, she pruned, fertilized, and kept weeds away from her trees and sold rooted cuttings to local nurseries. The fourth year, Maria had about 50 trees that were large enough to sell. She pruned and sprayed the trees, preparing them for the Christmas season. The first week of December, she cut the trees and sold them at the local farmers' market for $25 each. The fifth year, she sold 87 trees and replanted the trees with 2-gallon plants that she had propagated earlier. By her senior year in high school, Maria had an annual net income of over $6,000. Her agriculture class, part-time job, and SAE provided the knowledge, experience, and opportunity to profit from her interest in growing trees.

FIGURE 11–9

Hardwoods eventually become the climax vegetation.

loblolly, slash, and shortleaf. These species created canopies or overstories that suppressed the species of trees that were not tolerant of shade. However, beneath this canopy, hardwood species such as red oak and white oak grew. These species tolerate shade fairly well and will grow beneath the pine canopy. A characteristic of the southern pine is that it will mature and die at an earlier age than the oaks. When a pine dies, the space in the canopy may be filled with a hardwood such as oak. When the hardwoods create a canopy, little sunlight will penetrate. Because pines are intolerant of shade, they do not grow well under the oaks. The hardwoods eventually become the dominant species. This is called the climax vegetation because it is unlikely that other species will overtake the hardwoods (Figure 11–9). Succession is a very slow process that may take 200 years to reach the climax-vegetation stage. When the Europeans explored this region of the continent, they found a mixture of pine and hardwoods. The farther south they went, the more pine they found because the milder climate and sandy soil of the coastal plains favored the pines.

The Forest Ecosystem

Forests play an important part in our lives because they are a part of our ecosystem. An ecosystem is an area where all the living and nonliving features of the

FIGURE 11-10
Young, rapidly growing trees produce more oxygen than old trees.

area depend on each other. The forests of the world are instrumental in the support of life on the planet. Plants use carbon dioxide and give off oxygen in their life processes. Because animals use oxygen and give off carbon dioxide, plants support all animal life, including humans. Forests not only provide oxygen, but also aid in removing air pollution.

Young, rapidly growing trees produce more oxygen than older trees (Figure 11-10). As a tree produces new wood, carbon dioxide is removed from the air and oxygen is added. According to The American Forest and Paper Association, in the period of one year, an acre of healthy trees may produce 4,000 pounds of wood, use 5,889 pounds of carbon dioxide, and give off 4,280 pounds of oxygen. Since we have almost 500 million acres of trees, this represents a lot of oxygen that is put into the atmosphere.

Forests provide us with clean water (Figure 11-11). Roots of trees prevent water from running rapidly down a slope and instead cause the water to soak into the ground, where it accumulates. This recharges streams, lakes, and wells without adding the soil from runoff and helps keep the soil in place on slopes. In doing this, trees prevent lakes and streams from being muddy. Muddy lakes and streams prevent fish and other aquatic life from growing and lower our supply of clean drinking water.

FIGURE 11-11
Forest vegetation helps keep streams clean.

FIGURE 11-12
Loggers have to abide by regulations regarding timber harvest.

Forests provide habitat for animals and other plants. Many species of plant life live beneath the canopy of trees where they are sheltered from direct sunlight. These plants would not thrive in the open sun. Many species of wildlife inhabit the forests and depend on the trees for their existence. Different types of forests support different types of wildlife. Animals and birds prefer particular types of foliage for food or to make nests in. Without this habitat, certain species would become extinct in a short while. The federal government has established laws and regulations regarding the protection of animal and plant species in the forests. Loggers and timber companies have to abide by the regulations in the production and harvesting of timber (Figure 11–12).

The United States is one of the most beautiful countries in the world. In large part, this is because of our forests. Millions of acres of land are set aside by state and federal governments to protect the natural beauty of our forests. These national and state forests preserve a portion of the virgin forests that were here when the first settlers arrived. Today, these areas provide a valuable resource in terms of recreation (Figure 11–13). Hiking, fishing, hunting, and camping are favorite pastimes of many Americans and our forests make these hobbies possible.

FIGURE 11-13
Our forests provide us with areas for recreation.

One of the greatest contributions our forests make is the large amount of jobs associated with the forestry industry. Billions of board feet of lumber are grown, harvested, and processed each year. This requires a gigantic workforce to care for the trees, remove them from the forest, and convert the logs into finished products. This helps keep the nation's economy sound. It also supplies us with essential products that are used in almost all aspects of our lives.

Wood-Fiber Production

One of the largest segments of agriculture is the planting, management, and harvesting of trees. Americans at one time thought that the supply of timber was inexhaustible. Then around the turn of the twentieth century, people began to realize that it was possible to run out of forests to harvest if measures were not taken (Figure 11–14).

Many people also thought that the methods used to harvest timber were wasteful and led to the depletion of the soil. As a result, in 1891 Congress passed the Federal Forest Reserve Act. This legislation gave the president the power to set aside timberlands to be managed so that people in the future would have forests to use and enjoy. Later, large tracts of land

FIGURE 11-14
Around the turn of the century people realized that the forest would one day be depleted unless actions were taken. Clear cutting such as this was common all across the country.

were designated as federal timberlands. The National Forest Service was established to manage these lands. Legislation over the last 100 years has set aside land as being publicly owned. Some land is designated as national parks and is used only for recreational purposes. Other land was set aside as wilderness areas, where the environment is carefully preserved so that it will remain as close to its original natural state as possible. Still other public timberland is harvested by companies who buy the rights to cut the timber.

Forestry was once looked on as being a natural resource that could be utilized. However, in recent years the timber industry has become an agricultural enterprise. Trees are planted, cultivated, and harvested in much the same manner as other agricultural crops. This means that instead of being a resource that will one day be exhausted, the timber industry can be sustained year after year through the careful planting and cultivation of trees.

For many years, areas clear-cut of all or most of the trees have been seeded or set out with tree seedlings. This allows the forest to grow back in much the same manner that it originally did. Throughout the Northwest, forests have long been regenerated by scattering seeds on newly cut forest areas or by setting out small trees. For a forest to grow from seedlings to harvestable timber may take from 70 to 100 years. For this reason much of the timber industry has moved to the Southeast. In fact, the leading producer of wood is not Washington or Oregon, but Georgia. In this region, the hot humid weather allows the trees to grow in much less time. A carefully managed stand of timber can go from seedlings to harvestable trees in as little as 15 years. It must be remembered, however, that the timber from the Northwest provides higher-quality lumber for the construction industry. A large portion of the trees harvested in the South go for making paper or for plywood core material (Figure 11–15).

FIGURE 11–15
A large portion of the trees harvested in the South go into plywood.

Tree Farms

Tree farms are operations where trees are planted in rows and are given care throughout their growth period. At the proper time they are all harvested. New trees are then planted and the crop starts over again. There are tree farms in almost all states. The most are located in the South.

The process of tree farming begins in the nursery where superior seedlings are produced for tree plantations (Figure 11–16). Research goes into the selection and production of seeds from high-quality trees that are resistant to diseases and grow rapidly. Many of the seedlings are crosses between two species or strains of trees. For example, a longleaf pine may be crossed with a loblolly pine to produce a tree that is superior to either of the parent trees. This

FIGURE 11–16
Superior tree seedlings are produced in nurseries.

cross, known as a hybrid, benefits from a phenomenon known as heterosis or hybrid vigor. The hybrid grows more rapidly and is generally healthier than the purebred strains of trees.

After the trees are set out in large areas called plantations, they are managed using techniques that interrupt the process of succession. The underbrush may be cleared by using mechanical means, chemicals, or by prescribed burning. The use of fire in prescribed burning is aimed at destroying vegetation that competes with the trees for water and nutrients. The fire is carefully controlled to prevent damage to the trees and to prevent the fire from escaping.

Succession is also controlled by thinning the trees. As the trees grow and begin to compete with each other for sunlight and nutrients, some of the trees are harvested. Usually, the trees that are cut are those that are slowest growing or poorest in quality. The best trees are kept to grow. The trees harvested in the thinning process are used for fence posts, poles, or paper. As the trees grow and mature they are harvested and used for making paper, lumber, and other uses.

SUMMARY

Forestry is a large part of the agricultural industry. Trees are grown and harvested as a crop. If we continue to replenish the forests, we will always have wood to use to make the many products we enjoy. At the same time, we can enjoy the beauty of green forests.

CHAPTER REVIEW

Student Learning Activities

1. Take the side of a preservationist or a logger in a controversy over the proposed cutting of a tract of timber. Prepare your points and arguments, and debate someone with the opposite view.

2. Make a list of all of the products in your classroom that are made from trees. Do not forget the paper products. Compare your list with others in the class.

3. Select a particular type of tree that is grown for timber. Research the types of products that are made from the trees. What are the characteristics and uses of the lumber from the trees? Report to the class.

4. Visit a building-supply store and list all the different types of lumber sold. Identify the species of tree that produced the lumber and the region or country where the trees grew.

True/False

1. _____ One of the largest industries in agriculture is forestry.
2. _____ There are more trees in the United States today than there were 100 years ago.
3. _____ Wildlife greatly benefits from the forest industry.
4. _____ Plants use oxygen and give off carbon dioxide in their life processes.
5. _____ Forests aid in removing air pollution.
6. _____ Most tree farms are located in the North.
7. _____ Usually the trees that are cut are those that are growing best.
8. _____ Succession is controlled by thinning the trees.
9. _____ Preservation activists insist that as much of the old-growth timber as possible must be preserved.
10. _____ Succession is a very fast process.

Multiple Choice

1. At one time forestry was considered a
   a. conservation effort
   b. bad investment
   c. natural resource

2. The forestry industry stretches from coast to coast and encompasses around
   a. 500 million acres of trees
   b. 100 million acres of trees
   c. 5 million acres of trees

3. Each year there are close to _____ billion trees planted in the United States.
   a. one
   b. two
   c. five

4. The reason certain types of trees dominate a region is because of a natural process called
   a. deciduous
   b. succession
   c. overstory

5. An area where all living and nonliving features depend on each other is called an
   a. ecosystem
   b. oxygen system
   c. industry

6. The legislative act giving the president power to set aside timberlands is called the
   a. American Forest Act
   b. U.S. Forest Service Act
   c. Federal Forest Reserve Act

7. Young, rapidly growing trees produce more _____ than older trees.
   a. carbon dioxide
   b. oxygen
   c. pollution

8. One of the greatest contributions our forests make is the large amount of _____ associated with the forestry industry.
   a. jobs
   b. carbon dioxide
   c. hunting

9. The leading producer of wood is
   a. Washington
   b. Oregon
   c. Georgia

10. Operations where trees are planted in rows and are given care throughout their growth period are called
    a. nurseries
    b. loblollies
    c. tree farms

✺ Discussion

1. How does wildlife benefit from the forest industry?
2. What is the difference between hardwood and softwood timber?
3. What is a canopy?
4. What are the benefits of forests for animals?
5. What is the Federal Forest Reserve Act?
6. Which state is the leading producer of wood? Why?
7. Why is the harvesting of timber not really a threat to nature?
8. What is virgin timber?
9. What are the benefits of forests for humans?
10. Why did the cultivation of trees begin?

CHAPTER 12
Protecting the Environment

Student Objectives

When you have finished studying this chapter, you should be able to:

- Explain why everyone should be concerned with the environment.
- Discuss the importance of agriculture to a growing population.
- Tell how pesticides have the potential to cause harm to the environment.
- Explain how the threat of harm from pesticides is minimized.
- Discuss the difference between point and nonpoint pollution.
- Tell how agriculturists work to prevent water pollution.
- Explain the importance of wetlands.

Key Terms

environment
natural resources
renewable resources
carcinogen
Environmental Protection Agency (EPA)
hydrological cycle
aquifers
point source
nonpoint sources
wetlands

CHAPTER 12

Everyone should be concerned with protecting the environment. There is only a limited amount of space on our planet. Humans continue to multiply and the population grows. This means that there is less space for each person. At one time in history the population was so small that when conditions became bad, people could move to a new place. The places where people can move to a fresh environment are almost gone. Scientists predict that by the year 2050 there will be over 11 billion people in the world. That is about twice the current population. Most of you will be alive at that time and will have to cope with the large population.

As there will be less space for each of us, it becomes increasingly important that we care for the environment. The quality of life will be directly related to the conditions of the environment. Clean water to drink and clean air to breathe are essential. Most of the beauty we enjoy comes from the nature surrounding us (Figure 12-1). Few things are more pleasing than a green forest or a clear stream. The beauty of the surroundings will have to be protected from destruction. The more people living in the environment and the more people using our resources, the greater the likelihood that the environment will become polluted. Resources are those things we have available for our use. Natural resources occur in nature and they include

FIGURE 12-1

Most of the beauty we enjoy comes from the nature surrounding us.

FIGURE 12-2
With a huge and growing population, agriculturists will have to produce more food on less land.

water, air, mineral deposits, and soil. Usually when these resources are used up, they are gone forever. Renewable resources are resources that can be replaced, such as crops. These crops may include trees, flowers, and plants grown for food. In the future we will have to rely more heavily on renewable resources supplied by agriculture.

Not only will there be less space for all of us, but all of the people will need food to eat. Agricultural systems all over the world will have to supply the needs of over 11 billion people—and that number will continue to rise. With all these people needing space in which to live and work, less land will be available to grow food. Agriculturists will have to produce much more food on less land (Figure 12-2). Because of this, farming practices will have to become more intense, which means getting more production out of fewer acres of land. This will be a challenge to produce enough food and fiber to feed and clothe the world's population without wearing out the land or polluting the environment.

Agriculture always has and always will have an effect on the environment. In order to grow crops, forests have to be cleared, and less room is available for wildlife. To produce crops efficiently, fertilizers and pesticides have to be applied. The constant tillage of the soil can cause erosion (Figure 12-3). Animals that are raised in confinement produce a lot of manure that has to go somewhere. The answer

FIGURE 12-3
The constant tillage of the soil can cause erosion. Note the soil blown by the wind.

cannot be the elimination of agriculture. We must have food, shelter, and clothing. The answer is that we have to design our agricultural production methods so that they will protect the environment.

 Pesticides

Pesticides are substances that kill pests. Each year about 570 million pounds of pesticides are used to aid in the control of insects, weeds, and fungi. This may seem like a lot of chemicals, but you must realize that there are over 300 million acres of cropland in the United States. People have developed negative attitudes over the use of pesticides. Some view their use as being terribly destructive to the environment.

In the past, there have been problems with pesticides. Years ago a very popular insecticide was a chemical called DDT. This pesticide was very effective in killing insects and solved a lot of problems all over the world. However, DDT was very residual. This means that it stayed in the soil for a long time before it broke down. This caused problems with the food chain. Birds that ate insects affected by the insecticide retained DDT in their bodies. Water birds might eat fish that had consumed insects or other aquatic life affected by DDT. The DDT built up each time the birds ingested the chemical. Problems developed in the shells of the eggs the birds laid. The shells were so thin that the eggs would not hatch. This interrupted the life cycle of the birds. DDT also turned up in other forms of life. The chemical was passed up the food chain and all of the animals in the chain were affected. Because of these problems, DDT and chemicals like it have been banned. Modern pesticides break down very rapidly and pose much less danger than the older pesticides.

Sometimes, problems with agricultural chemicals are blown out of proportion. For example, in 1989 the media began a series of stories on the chemical daminozide, known by its trade name, Alar. This chemical acts as a growth regulator that is used on apples to prevent the apples from falling off the tree too early. This helps the apples to grow into a better shape. Also it prolongs the storage life. According to the reports, the chemical proved to be a **carcinogen** (a substance that causes cancer) when fed to laboratory rats (Figure 12–4). As a result, near panic ensued among consumers. Apples and apple products were taken off the market. Tons of apples were dumped as garbage and the fruit was banned from school cafeterias. The apple industry suffered tremendous losses. The money earned from the sale of apples fell by more than $100 million. In addition, millions of dollars of income were lost from all the support

FIGURE 12–4

The chemical Alar was reported to be a carcinogen to laboratory animals.

FIGURE 12-5
A person could not eat enough apples in a lifetime to receive the amount of Alar given to the lab animals.

services of the apple industry. Many apple producers reached the brink of bankruptcy. This example points to the concern of the public about the safety of food.

Further investigation showed little justification for the panic. In the study, laboratory rats had been given enormously high amounts of Alar before they developed tumors. At much lower rates the chemical had no effect at all. According to the American Council on Science and Health, a person would have to eat the equivalent of 28,000 pounds of apples per day for 10 years to equal the amount of Alar fed to laboratory animals before they developed tumors. The amount of Alar that was on the apples was almost unmeasurable. In no way could a human eat enough apples in a lifetime to consume enough Alar to cause problems (Figure 12–5). Yet in spite of this fact, the chemical was pulled off the market.

The use of pesticides are necessary if we are to provide the population with high-quality food. Without them both the quantity and the quality of food would be lower. Produce that is damaged by insects or disease will be of such low quality that consumers will be reluctant to accept it. Disease causes bad spots on fruits and vegetables. Fruits and vegetables that have been partially eaten by insects will not only look bad and taste bad but will also spoil more easily.

PROTECTING THE ENVIRONMENT

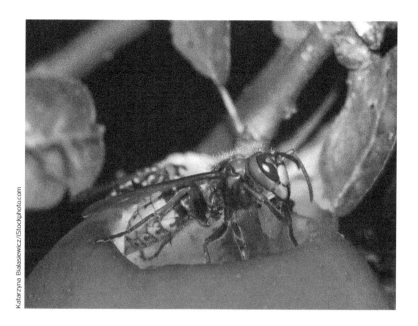

FIGURE 12-6
Pesticides are necessary to provide wholesome food. This fruit is not edible because of pest damage.

Used properly, modern pesticides are safe to use. The United States Department of Agriculture (USDA) and the Environmental Protection Agency (EPA) have strict regulations regarding the development and use of pesticides. Before a pesticide can be released for use, it must undergo years of rigorous testing to ensure it is safe. It must not only pose no threat to humans but must not be harmful to the environment. Modern pesticides are not only necessary but safe if used properly (Figure 12–6).

Water Pollution

One of the most precious natural resources we have is our water. This compound is essential for life. Without it, humans would die within a few days. We depend on it to sustain the lives of the plants and animals we grow for our food supply (Figure 12–7). Most of our industries require large amounts of water, and so does much of our recreation. Fishing, boating, and swimming require a huge amount of clean water. We have only a certain amount of water. Water is used and recycled over and over again through a process known as the hydrological cycle (Figure 12–8). Water evaporates, condenses, and returns to the earth as rainfall.

FIGURE 12-7
We depend on clean water to sustain life.

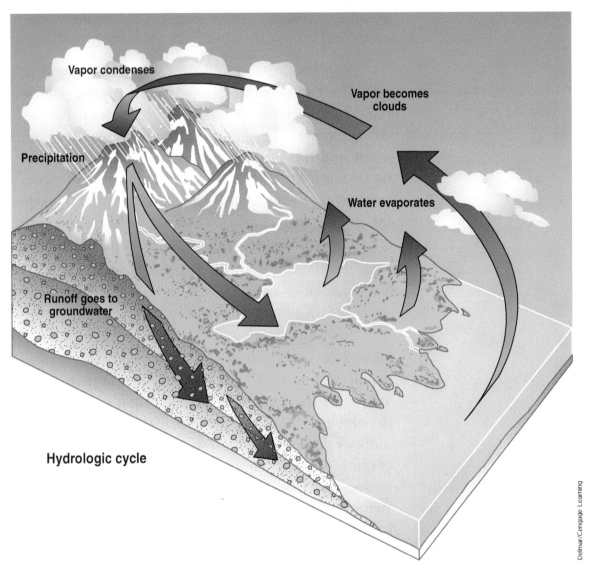

FIGURE 12-8
Through the hydrological cycle, water is constantly recycled.

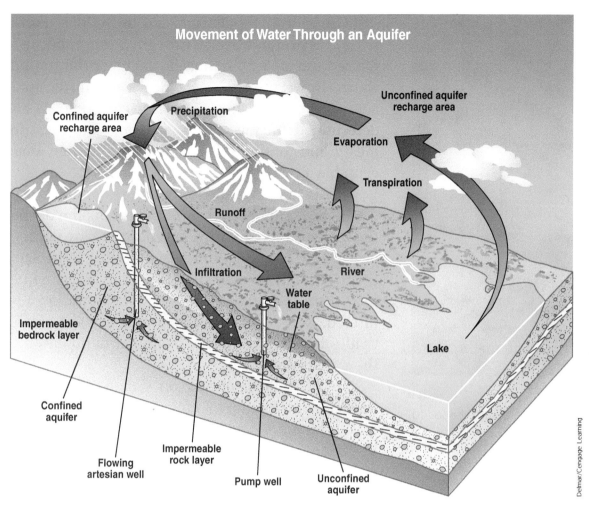

FIGURE 12-9
Aquifers are formations in sand and rock that hold water.

Rainfall replenishes two types of water storage—surface water and groundwater. Surface water is the water that is in streams, lakes, and the ocean. When rain falls, the amount that is not absorbed into the ground flows downhill into streams. Small streams flow into larger streams and the large streams eventually flow into the oceans. A few years ago, surface waters in this country faced serious pollution problems. Many streams and lakes were so polluted that very little life could exist there.

The other type of water is called groundwater. Groundwater is the water that is stored under the surface of the ground. When rain hits the ground, some runs off and some is absorbed into the ground and accumulates in storage areas called aquifers. Aquifers are natural formations underground in sand, rock, or gravel that hold water (Figure 12-9). Wells are drilled into aquifers to retrieve water. A large portion of our drinking water comes from aquifers.

FIGURE 12-10
The EPA requires that water returned to a stream be clean.

Groundwater pollution can be serious. Fortunately, as water travels through the ground, it is filtered and most pollutants do not reach the groundwater. However, where accidental chemical spills occur, problems can happen. However, the EPA has determined that fewer than 1 percent of the rural domestic and community water systems contain unacceptable levels of pesticides.

Areas that have shallow aquifers or coarse, sandy soil are in more danger of pollution. The deeper the water supply and the heavier the clay content of the soil, the safer the water is. This is because the depth and the clay slow down the movement of the water. This gives more time for the water to be filtered. New regulations governing the use of pesticides have lessened the likelihood of pollution of the groundwater.

In 1970, President Nixon signed into law the National Environmental Policy Act, which led to the establishment of the EPA. This agency was charged by Congress to "protect the nation's land, air, and water systems." Policies, laws, and regulations have been and continue to be established to prevent pollution and to clean up pollution that has already occurred. Since 1970, a lot of progress has been made in cleaning up the air and water. Many bodies of water that were once considered dead now have been brought back to life by cleaning up the water and stopping pollution.

Pollution comes about as a result of contaminants entering the water in two main ways: point and nonpoint sources. A point source means pollution from a specific place. For instance, a factory that dumps waste into a river is a point source of pollution because the contaminants can be traced to a single point. The EPA has made tremendous strides in halting point-source pollution. Industries have been required by law to stop dumping or allowing contaminants to enter waterways. Water used in industry and returned to streams or lakes is carefully monitored to ensure that it contains no polluting substances (Figure 12-10).

Pollution from nonpoint sources is more difficult to deal with. Nonpoint means that the pollution does not come from a single point but from a wide area that is difficult to identify. The pollution is usually from a number of sources that collectively affect the environment. Agriculture is the largest origin of nonpoint pollution in the United States. This is due in part to the enormous size of the agricultural industry, where millions of acres of land are in production. Each year, this land receives treatment in the form of plowing, cultivating, storage of animal wastes, and the application of pesticides. As a result, agricultural pollution comes from several sources: fertilizers, pesticides, animal wastes, and soil erosion. Recently, laws have been passed to help protect the environment from these pollutants.

Soil Erosion

Our largest environmental problem stemming from agriculture is soil erosion (Figure 12–11). Erosion is the wearing away of the soil by wind or water. This problem is twofold: the loss of topsoil and the pollution of surface water by the displaced soil.

Research indicates that the sediments from soil cause around $6 billion in damage each year. Almost all improved land in the United States has a system of drainage ditches that carries off excess water.

FIGURE 12–11

Soil erosion is our most serious pollution problem stemming from agriculture.

Eroded soil clogs these waterways and has to be periodically removed. Similarly, navigable waterways can become so clogged that they have to be dredged out in order for the vessels to pass through. This is a tremendously expensive process that takes a lot of time and resources.

Another serious problem caused by sediment pollution is the loss of fish and other aquatic life in lakes and streams. Some species cannot tolerate muddy water and die from lack of oxygen or lack of clean water (Figure 12–12).

As the agricultural industry has developed, methods of protecting the soil from erosion have developed also. Intensive efforts began back in the

COMMUNITY DEVELOPMENT IN FFA

In every community across the nation there is a need for improvement. The FFA program encourages students to identify community needs and take action. Through the years, community programs have developed active, experienced, and knowledgeable leaders by educating and involving students. Cities, towns, and communities across America have benefited from these projects. Projects have created and expanded community services and improved the quality of life. Community development is one of the major areas addressed in a local FFA chapter program of activities. To make a meaningful impact upon the community, the whole chapter should be involved in the planning and action. The first step is to select a chairperson and committee. The chairperson will lead the chapter in the planning, initiation, and public relations for the effort. The committee will identify needs through questionnaires, attending public meetings, or surveys of the community. After collecting the information, the committee will assess the information and identify a community project. Identification of resource persons and community groups for support of the activity is important. The involvement of groups such as FFA Alumni, Chamber of Commerce, and civic clubs can not only provide financial support, materials, and equipment but offer expertise in needed areas. Cooperative efforts with other community groups will increase the success of the project. A plan of action is necessary to organize the phases of the project and its completion. A step-by-step project plan, work schedule, budget, and available

FIGURE 12-12
Some species of aquatic life cannot tolerate muddy water and die from the lack of oxygen.

Recycling is just one of many ways FFA members can assist in community development by helping keep the environment clean. *Courtesy of Blane Marable, Georgia Agricultural Education.*

resources should be included in the plan of action. After developing the plan of action, it is time to start the project. Involving all FFA members and interested community groups will result in a successfully completed project. Publication of community development activities through newspapers, radio, and television will increase community awareness and support. After the project is completed, a community-wide meeting, such as an FFA banquet, should recognize those groups and individuals who participated. You can make a real contribution to your community!

1930s with the creation of the Soil Conservation Service (SCS). The mission of this agency was to help prevent the loss of soil and to stop pollution caused by erosion. The SCS has made tremendous strides in the control of erosion. The name was changed in the early 1990s to the Natural Resources Conservation Service (NRCS). Modern efforts are aimed at keeping as much vegetation on the ground as possible. Plants hold the soil together by creating a support system for the soil particles. Roots penetrate the soil and through their fibrous structure hold the soil in place. Many crops are now grown with a bare minimum of plowing (Figure 12-13). This not only allows more plant roots and plant cover to hold the soil but also slows the flow of water over the land. In some areas, crop residues were plowed under but now are left until spring. The new crop is planted in the stubble left from the previous crop.

New varieties of grasses and other plant covers are also being developed. Bare hills are planted in grasses that grow quickly, have an expansive root system, and do well on eroded soils.

FIGURE 12-13
In some areas the soil is tilled as little as possible to prevent soil erosion. Note the residue from last year's crop that is still between the cotton plants.

Nitrate Pollution

Water is also polluted by agricultural production through the use of commercial fertilizers and animal wastes. Animals that are raised in confinement produce a lot of manure in a concentrated area (Figure 12-14). This material is generally used on fields as a fertilizer. Within both fertilizers and manures are plant nutrients called nitrates. In surface water, a major problem is that nitrates cause plant life in streams and lakes to flourish. This may not seem like a serious problem, but too much plant matter in the water can cause trouble. Clogged waterways are difficult for boats to maneuver through and interfere with recreational activities. Also, when the plant material dies and decays, oxygen is taken from the water. When a large amount of vegetation decays, much of the oxygen can be removed from the water and the fish may die.

FIGURE 12-14
Animals in a confinement operation produce a lot of manure. Most of it is spread on fields as fertilizer.

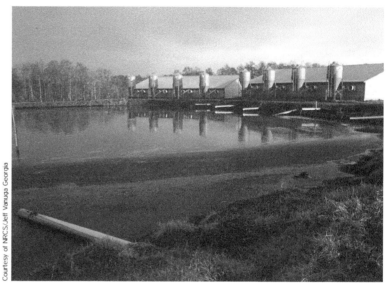

FIGURE 12-15
A lagoon is a pond of water into which waste material like manure is directed. The material decomposes in the water.

Two practices can help control the problem. The first is to control the runoff from rainwater. Diversion ditches are cut into the area surrounding the livestock facility to prevent runoff containing waste materials from getting into streams. The drainage ditches divert the runoff into holding areas where the risk from runoff is at a minimum.

The second practice is to use a lagoon as a holding tank for the manure. Basically a lagoon is a pond of water into which the waste is directed. Here the material is decomposed by anaerobic bacteria (Figure 12-15). These bacteria live beneath the surface of the water in the lagoon and thrive without oxygen. Anaerobic bacteria break down the waste material and convert the nitrogen into the gaseous state. At the proper stage of decomposition, the material is pumped from the lagoon and spread on the fields.

Wetlands

Many areas in our country are known as **wetlands**. A wetland is an area where the ground is naturally saturated for much of the year. These areas were once thought to be wastelands that served no good purpose and should be drained to be productive. It is estimated that over half of the wetlands of the United States have been destroyed for construction or for agricultural purposes.

Scientists now know that wetlands play several essential roles in the ecosystem of an area (Figure 12–16). These swampy areas help to filter water before it reaches open areas that store surface water. Earlier in this chapter, problems with chemicals, sediments, and other pollutants in runoff water were discussed. Wetlands catch runoff water before it reaches deeper bodies of surface water. The vegetation removes nutrients and other pollutants from the water, and as a result the open areas of surface water are much cleaner. Wetlands also act as a sponge or storage area that collects and absorbs floodwater before it can damage crops, homes, or other property.

FIGURE 12–16

Scientists now know that wetlands play several essential roles in the ecosystem of an area.

FIGURE 12-17
Some species of plants will only grow in wetlands.

Wetlands are critical to the life cycle of many different plants and animals. They may provide water for drinking, food, or shelter for a variety of animals in the area. Also, some plant species will only grow in areas where the ground remains saturated for most of the growing season (Figure 12-17). If wetlands are destroyed, many of these plants may be lost forever.

The EPA has established regulations that protect wetlands by restricting their use. Incentives are offered to landowners to promote the preservation of wetland areas. Permits must be obtained from the government when wetlands are to be disturbed. Through these measures valuable wetlands will be protected and in turn the whole environment will be enhanced.

SUMMARY

Agriculturists will continue to help protect the environment. It is in their best interest and in the public interest. We must have food and fiber. This can only come about through agriculture. Scientists and other concerned individuals will continue to find new and better ways of producing food and fiber, and protecting the environment.

CHAPTER REVIEW

Student Learning Activities

1. Survey your community and decide if the environment in your area is being cared for. If so, list the measures taken to preserve the environment. If you feel there is room for improvement, outline steps that can be taken to solve the problems.
2. Locate an area where there has been soil erosion. Outline a plan that can help prevent more loss of the soil. Be sure to list the major factors that contributed to the erosion.
3. Go to the library and locate a story of a pollution problem that was solved. Report to the class on your findings.
4. Talk to several elderly people who have lived for many years in your area. Get their perspective on how the environment has changed. Do they think the environment is better or worse? Report to the class.

True/False

1. _____ The deeper the water supply and the heavier the clay content of the soil, the more likely the water will be polluted.
2. _____ The National Environmental Policy Act led to the establishment of the Environmental Protection Agency.
3. _____ Since 1970, a lot of progress has been made in cleaning the air and water.
4. _____ Erosion is the wearing away of soil by wind or water.
5. _____ Research indicates that the sediments from soil cause approximately $6 billion in damage per year.
6. _____ Diversion ditches are ponds of water where waste is directed.
7. _____ Over half of the wetlands in the United States have been destroyed for construction or agricultural purposes.
8. _____ The Soil Conservation Service (now NRCS) is concerned with minimizing pesticides in soil.
9. _____ All crops are now grown with extensive plowing.
10. _____ Nitrates cause plant life in streams and lakes to flourish.

Multiple Choice

1. One of the most precious natural resources we have is
   a. crops
   b. soil
   c. trees

2. Rainfall replenishes two types of water storage: groundwater and
   a. surface water
   b. lagoons
   c. stored water

3. The water in streams, lakes, and oceans is
   a. groundwater
   b. surface water
   c. polluted water

4. Formations in sand, rock, or gravel that hold water are
   a. wells
   b. wetlands
   c. aquifers

5. Pollution comes about as a result of contaminants entering the water in two main ways:
   a. point and nonpoint sources
   b. rain and snow
   c. fish and birds

6. The largest origin of nonpoint pollution in the United States is
   a. rain
   b. industry
   c. agriculture

7. Our largest environmental problem stemming from agriculture is that of
   a. soil erosion
   b. dairy farming
   c. excess pesticides

8. Permits must be obtained from the government when these are disturbed:
   a. wells
   b. wetlands
   c. aquifers

9. Years of rigorous testing are required to ensure the safety of
   a. crops
   b. wells
   c. pesticides

10. Resources that can be replaced are
    a. renewable
    b. residual
    c. hydrological

Discussion

1. What are natural resources?
2. What are renewable resources?
3. Why must farming practices become more intense?
4. What is the difference between modern and older pesticides?
5. What are some uses for water?
6. Where does agricultural pollution come from?
7. What is soil erosion?
8. Why is sediment pollution such a large problem?
9. Why is it better that crops are grown with a bare minimum of plowing?
10. Why are wetlands so important?

CHAPTER 13

Organic Agriculture

Student Objectives

When you have finished studying this chapter, you should be able to:
- Discuss the meaning of the term organically raised product.
- Explain the concerns some people have about our food supply.
- List ways organic producers fertilize their crops.
- List ways organic producers control insect pests.
- Analyze the controversial issues surrounding organically raised products.

Key Terms

organic
commercial fertilizers
cover crop
composting
humus
soil amendments
fish slurry
organic fertilizer
crop rotation
border plants
predator insects
natural toxins

We live in a very affluent society in which we buy almost everything that we consume, and food is no exception. Many years ago, just about everyone grew at least some of the food they ate. If people did not farm for a living, they at least grew a garden and kept cows for milk and hens for eggs. Because they grew most of the food eaten, people knew where their food came from and how it was produced, and so they felt the food supply was safe.

In a society in which few people grow food, most are not aware of where their food comes from, or how it is produced. Food is produced in huge quantities and most often the production takes place many miles from where it is consumed. The efficiency of American producers is due to the use of technology, which has led to the use of chemical fertilizers and pesticides. This has allowed us to produce enough food to feed all of our citizens and much of the world's people.

Advances in laboratory techniques have allowed scientists to identify minute quantities of pesticide and chemical residues on foods. Just a few years ago, these small amounts of residue could not be detected. Regulations by the USDA (U.S. Department of Agriculture) specify the maximum amount of residues allowed on produce sold in this country. Many people feel that any chemical residue on food is too much and that any food produced using chemicals such as pesticides and commercial fertilizer is not healthy (Figure 13–1). They reason that for food to be the most healthy and wholesome, it should be produced using only "natural" inputs.

These concerns have led to a new agricultural industry called organic agriculture. In this system food and fiber are produced using only materials that are "organic" in origin. Organic materials are generally considered to be composed of substances that occur only in nature and did not result from a manufacturing process.

The meaning of "organically produced foods" has now been defined by the USDA. However, there is not a short definition of organically grown foods.

FIGURE 13-1
Many people consider any food that has been sprayed by pesticides to be unhealthy.

FIGURE 13-2
Markets for organic foods have grown steadily.

There are many rules and regulations outlining what can and cannot be used in organic production. For example, many types of pesticides are prohibited, as well as certain types of fertilizers. The USDA also lists substances that can be used. For example, manure or animal waste can be used if it is properly composted. This means that the manure must be stored for a length of time so chemical reactions that produce heat may occur. The heat kills odors, weed seed, and many types of bacteria. If the qualifications are met, the product may be labeled as "organically grown." Many other states have similar organizations.

Organic farming has been growing steadily for over 30 years (Figure 13-2). Today there are almost 1½ million acres of organically produced crops in the United States. This acreage increases each year.

The Production Process

Methods used by organic producers are a lot different from those used by conventional producers. A majority of all the food and fiber products produced in the United States is grown using commercial fertilizers and pest controls. Commercial fertilizers are produced in factories from chemicals such as nitrogen, phosphorus, and potassium. Also, much of the beef is grown using growth stimulants or hormones to improve the growth efficiency of the animals. Organic producers use a different production practice. They use natural methods of fertilizing and pest control.

Fertilizers

Producers of organically grown foods and fiber use a variety of methods to provide plants with nutrients. The idea is to use only materials occurring in nature and not to use fertilizers made purely from chemicals. One of the basic methods is to use **cover crop** plants such as annual winter grasses or legumes that are planted each year when crops are harvested. The cover crops grow during the winter and early spring until the time when new crops are to be planted. They are then turned under before the next crop is planted to provide organic material for the new plants (Figure 13–3). The cover crop decays and through the natural process of decomposition releases nutrients into the soil.

In addition, cover plants such as vetch and clover are legumes that have nodules on the roots that host nitrogen-fixing bacteria. These bacteria take nitrogen from the air and convert it to a form of nitrogen

FIGURE 13-3

Organic producers often plant cover crops to add organic material to the soil. Crimson clover is sometimes used as a cover crop.

ORGANIC AGRICULTURE

FIGURE 13-4
Organic materials are composted to produce natural fertilizer.

usable by plants. The process of nitrogen fixation greatly improves the producing ability of the soil.

Another method used by organic producers is that of composting. Materials such as leaves, grass clippings, and various other plant parts are put into a bin designed to speed the decomposition process. As plant material decays it is converted into a substance called humus. Humus contains nitrogen and also increases the water- and nutrient-holding capacity of the soil. Other organic materials such as peat moss or decayed sawdust (called soil amendments) are added to the soil. Other amendments such as lime or gypsum may be used to adjust the pH level of the soil to suit the particular crop being grown.

There are several sources where producers of organically grown products can obtain materials for use as fertilizers (Figure 13-4). One is the fishing industry. When fish are harvested and dressed, a large portion of the fish carcass remains after the edible portions are removed. This combination of heads, skin, entrails, fins, and bones is ground and blended into a slurry that is used as a nutrient-rich fertilizer. Fish slurry is high in nitrogen content and also contains phosphorous and potassium. These three nutrients are all major nutrients required by plants.

Perhaps the most important source of organic fertilizer is animal manure. All areas of the animal industry produce millions of tons of fecal matter each year that has to be disposed of properly. Because manure is very high in nutrient content it is a valuable fertilizer. The dual purpose of getting rid of waste and fertilizing crops is then achieved. Unless the proper steps are taken, major problems may exist with using manure. One problem is that manure causes quite an odor problem when it is spread onto a field near where people live.

Producers who grow crops near a residential area and fertilize with raw manure that has not been properly composted can come under heavy criticism from people living in the area. A second problem with using manure is that the fecal matter contains high levels of *E. coli* and other forms of bacteria that may be harmful to humans.

A third problem with using manure is that most raw manure contains a considerable amount of weed seed. When grain is grown, almost always some amount of weeds will grow up and mature in the field alongside the crop. When the crop is harvested, the seeds from the weeds are also harvested. Many weed seeds are too small to separate out when going through the processing with the grain. Because most of the weed seeds are not harmful to livestock, they are left in the grain as part of the processed livestock feed. The weed seeds are then ingested by the animal, go through the animal's digestive system, and are passed out with the feces. When the feces are used as manure fertilizer, the weed seeds are planted and may germinate in the soil along with the crop.

Using proper steps in the process of composting can eliminate all three of these problems (Figure 13–5). When the manure is composted, it is piled together and allowed to age for at least two months prior to use. If properly composted, the internal temperature of the manure should reach 130 to 140 degrees F. This degree of heat should be enough to kill most harmful bacteria and weed seed. In addition, the composting process helps ease problems with odor.

ORGANIC AGRICULTURE

FIGURE 13-5
If manure is properly composted, problems with odor, weeds, and bacteria can be eliminated. Note the steam rising from the composted manure.

 ## Insect Control

One of the greatest problems facing organic producers is that of controlling insects. When plants are growing and weather conditions permit, populations of insects can build up rapidly and destroy a crop very quickly (Figure 13-6). Conventional producers use a variety of chemical weapons to control insects. However, organic producers are against using insecticides to kill pests and use more natural methods such as crop rotation to help in insect control.

The principle behind crop rotation as an insect control method is that many insect pests attack only particular plants. By planting fields with different crops each year, populations of insects have less chance of growing because the plants they feed on are not grown two years in a row. For example, hornworms may attack tomatoes but may not bother squash. Planting squash in a field that had tomatoes the previous year deprives the hornworms of food. This helps break up the insects' life cycle.

Another method is planting border plants around the outside of the field where the crop is grown. These borders serve to provide an alternative crop for the insects. The idea is that the insects attack the plants around the border and not the crop.

FIGURE 13-6
Populations of insects can build rapidly and destroy crops.

FIGURE 13-7
Predator insects such as the lady beetle can help control harmful insects.

Border plants are usually perennials that do not have to be planted each year and must be plants that insects prefer to the crop plants.

One of the most effective weapons organic producers have in controlling insects is the use of predator insects. In nature many insects live by eating other insects. There is a high population of predator insects such as praying mantises, lacewings, lady beetles, and parasitic wasps (Figure 13–7). Many of the predator insects are grown commercially for the purpose of selling to organic producers. Modern garden supply stores offer predatory insects for sale through their retail catalogs. As soon as insect pests appear, the organic producer releases large numbers of the predatory insects to eat the harmful insects.

Specific predators are used for specific pests. For example, lacewings feed only on thrips and aphids, so if these insects are a problem the producer introduces a large number of lacewings. The larvae of parasitic wasps eat a variety of insects and are used as a method of killing more than one type of insect pest. The wasp lays eggs on or in the harmful insect and, on hatching, the larvae feed on the insect.

Organic Animal Agriculture

Although most organic production is centered around plants, animals are also grown and sold organically. Most often, livestock products labeled as organically produced applies to animals that have been raised without pesticides to control internal and external parasites. Also the animals must not have been exposed to the use of growth stimulants or hormones. The labeling also extends to the use of irradiation for preserving meats. Although the USDA allows this method of preservation, meats treated with radiation may not be labeled as organically produced. Animals are generally raised in open-air, noncrowded conditions, which include the humane treatment of the animals (Figure 13–8).

FIGURE 13-8
To be certified as organically produced, animals generally have to be raised in open-air, noncrowded conditions.

Sometimes the certification may require that the animal be fed organically produced feed and raised by other strict nutritional requirements. For example, the California Certified Organic Farmers prohibit the following:

1. Nonorganic feed
2. Feed containing plastic pellets, urea, synthetic preservatives, or other synthetic materials
3. Feed containing antibiotics, hormones, drugs, or synthetic growth promoters
4. Intentional refeeding of manure
5. Maintaining animals on poorly balanced diets, including raising anemic animals
6. Feed containing animal by-products for ruminants
7. Feed that has been solvent extracted

Only products from animals that meet all the production requirements may be labeled as organically produced. This includes products such as wool and hair, meat, milk and milk products, eggs, and any other animal-derived product.

Criticisms of Organic Production

Since the concept of organic production began, controversy has surrounded the issue. Critics maintain that the public is misinformed about the safety and nutritional value of organically grown products. They

FIGURE 13-9
Research has failed to show that organically produced foods are more nutritious or healthier than conventionally produced foods.

point out that these products are much more expensive than conventionally produced food and fiber and are generally of poorer quality. Critics point out that a lot of research has been done comparing the safety and nutritional value of organic foods to conventionally grown foods. To date, no definite evidence has shown that organically produced foods are safer or more nutritious than conventionally produced food (Figure 13-9).

In fact, there are several disadvantages. One of the main reasons for organically grown food is to avoid the toxins from the trace amounts of insecticides that could possibly be on regularly produced fruits and vegetables. Plants have a self-defense mechanism that produces natural toxins to help repel insects. If attacked by pests, the plants release these substances because of the stress caused by the insects or disease. These natural toxins can be many times the amount produced by plants that were protected by pesticides.

It is difficult to produce fruits and vegetables that have the quality we are used to without the use of chemical pesticides. Natural methods of controlling pests can be effective, but when used alone cannot offer the protection that will ensure that the food does not contain insect damage. This damage on fruits and vegetables provides a natural harbor for bacteria and other organisms that cause the food to

FIGURE 13-10
Insect damage on fruits can provide a harbor for bacteria and other harmful organisms.

spoil (Figure 13-10). Most scientists agree that the greatest danger we have from our food supply is that of food spoilage and the ingestion of bacteria, not the presence of pesticides.

Another problem with organically produced food is that often manure is used as fertilizer. Unless manure has been treated, it contains bacteria that can be harmful to humans. Critics point out that contact with these bacteria greatly offsets any advantage offered by using manure. Also, they point out that nitrogen, phosphorus, and potassium, which are the major nutrients needed by plants, are the same whether they come from organic or inorganic sources.

Critics argue that organic production requires more land and more labor to produce the same amount of products grown using conventional means. In fact, some scholars argue that intensive farming methods are much more "environmentally friendly" than organic production. Projections are that by around the middle of the twenty-first century, the world's population will double. This means that the world's production of food and fiber will have to double through either of two methods. One is to clear new land and increase the number of acres under cultivation. The second is to grow more food and fiber on the land currently under production. The concern raised over organic production is that its practice will necessitate the clearing of more land.

FIGURE 13-11
Critics contend that not enough food can be produced using organic means to feed the world's population.

As more land is cleared, the loss of wildlife habitat will increase. Most of the new land being cleared is through the "slash and burn" methods used in tropical areas. Much of the land cleared is poor quality and highly susceptible to erosion. However, through the use of conventional, more intensive farming, less new land will have to be put into production.

The fact that organic production relies on the use of organic materials is also a weakness of the system, especially if this were the main method of producing our food and fiber. If all the animal manure and all the beneficial organic materials available were used, there would not begin to be enough to fertilize all the cropland. Obviously, fertilizer inputs will be needed, and the argument is made that organic sources alone can supply only a small fraction of what is needed (Figure 13-11).

SUMMARY

Despite the controversy over the issue, the production and consumption of organically grown products continue to grow. This relatively new branch of the agricultural industry is here to stay and will receive increasing attention in the years to come. As with all other aspects of the market, it is driven by consumer demand, and as that demand increases, production will also increase. The challenge will be to keep consumers well informed on research findings related to organically raised products.

CHAPTER REVIEW

Student Learning Activities

1. Go to the grocery store and make a list of all the products you can find that were organically produced.
2. Research and prepare a five-minute speech giving your position on organic agriculture.

3. Compare organically produced fruit (like an apple) with conventionally grown fruit of the same type and variety. List the differences in cost, appearance, and taste.

True/False

1. _____ Organic agriculture uses only materials that originated from living material.
2. _____ *E. coli* is one of the problems when using manure as a fertilizer.
3. _____ Controlling insects is not a very big problem for organic producers.
4. _____ When irradiation is used, the meat can still be considered organic.
5. _____ Most scientists agree that the greatest danger we have from our food supply is that of food spoilage and the ingestion of bacteria.
6. _____ If only organic farming is used in the future, there will not be enough organic material to fertilize all the crops planted.
7. _____ Clearing land for organic farming will have no impact on wildlife.
8. _____ Nitrogen-fixing bacteria take nitrogen from the air and convert it to a usable form for the plant.
9. _____ A problem with using animal manure for fertilization is that it contains weed seed.
10. _____ The least effective weapon organic producers have in controlling insects is the use of predator insects.

Multiple Choice

1. Today there are almost _____ acres of organically produced crops in the United States.
   a. 2 million
   b. 1 billion
   c. 1½ million
   d. 500 thousand

2. Which is not an example of an organic fertilizer?
   a. fish slurry
   b. compost
   c. manure
   d. 10-10-10

3. Surrounding a field with a different crop is an example of
   a. composting
   b. using border plants
   c. using a cover crop
   d. edging

4. Which is an organically produced product from an animal?
   a. wool
   b. meat
   c. milk
   d. all of the above

5. At what internal temperature should manure be heated to kill harmful bacteria?
   a. 100 to 120 degrees F
   b. 220 to 230 degrees F
   c. 180 to 190 degrees F
   d. 130 to 140 degrees F

6. What does USDA stand for?
   a. United States Demands Action
   b. United States Department of Agriculture
   c. Use Safety Doing Agriculture
   d. United States Department of Action

7. Most of the land being cleared for agriculture is cleared with the _____ method.
   a. slash and burn
   b. clear-cutting
   c. seed tree
   d. shelterwood

8. Which is not an example of a cover crop?
   a. legumes
   b. winter grasses
   c. trees
   d. clover

9. Border plants are usually
   a. biennial
   b. annuals
   c. perennials
   d. none of the above

10. Which is an example of a predator insect?
    a. lacewings
    b. lady beetles
    c. parasitic wasps
    d. all of the above

✹ Discussion

1. Discuss why people are in favor of organic agriculture.
2. Discuss the use of crop rotation.
3. Explain how lacewings are helpful predator insects.
4. List five of the seven items California Certified Organic Farmers prohibit when producing organic animals.
5. Discuss some of the criticism of organic production.

CHAPTER 14
The Livestock Industry

Student Objectives

When you have finished studying this chapter, you should be able to:

- Discuss the extent of the U.S. meat industry.
- Discuss the different phases of the beef industry.
- Explain how pork is produced in the United States.
- Discuss the extent of the sheep industry.
- Describe the types of and uses for horses in this country.
- Describe the small-animal industry.

Key Terms

per capita consumption
beef
breed
carcass
crossbreeds
purebred
ration
pork

FIGURE 14-1
Much of the land in the United States that is not suitable for row cropping is suitable for growing animals.

The livestock industry in the United States is a tremendously large industry. Animals are raised in all 50 states. They are used for meat, milk, and dairy products (see Chapter 15), clothing (wool and leather), and recreation (pets and horses). For as long as there has been a country, Americans have raised large numbers of livestock. Our country is well suited for growing livestock. Much of the open land in this country that is not suitable for growing row crops is suitable for growing animals (Figure 14-1).

The Meat Industry

The average person in this country consumes almost 200 pounds of red meat and poultry each year. Very few nations in the world even come close to us in the per capita consumption of meat (Figure 14-2).

Some health authorities have expressed concern over the consumption of so much meat. Research has shown that a substance called cholesterol in the fat of meat can cause heart problems. Most health experts agree that a moderate intake of meat is necessary for a healthy diet. Meat is among the most nutritionally complete foods that humans consume. Foods from animals supply about 88 percent of the vitamin B_{12} in our diets because this nutrient is very difficult to obtain from plant sources. In addition,

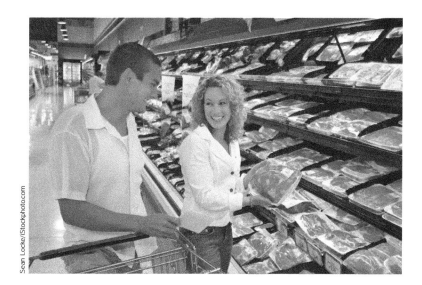

FIGURE 14-2
Very few nations in the world eat as much meat as Americans.

meats and animal products provide 67 percent of the riboflavin, 65 percent of the protein and phosphorus, 57 percent of the vitamin B_6, 48 percent of the fat, 43 percent of the niacin, 42 percent of the vitamin A, 37 percent of the iron, 36 percent of the thiamin, and 35 percent of the magnesium in our diets. Problems with cholesterol may be avoided by exercise and eating a reasonable amount of meat.

The Beef Industry

Beef is by far the largest segment of the meat industry. In the United States, there are over 800,000 producers of beef. Over three-quarters of all of the money earned from the meat industry comes from beef. The average size of beef herds in the United States is about 100 animals.

In the United States there are over 40 different breeds and many different combinations or crosses of these breeds. A breed is a group of animals with a common ancestry and like characteristics (Figure 14–3). The offspring of these animals look like their parents. Livestock producers choose breeds based on several factors. The market where the animals will be sold, the environmental conditions the animals will be produced in, and the personal likes and dislikes of the individual producer may be considered.

FIGURE 14–3

A breed is a group of animals with a common ancestry and similar characteristics. These cattle are Belted Galloways.

Some breeds are large and produce a large carcass; some mature at a smaller size and produce a smaller carcass. A carcass is that part of a meat animal remaining after slaughter, and the hide, head, feet, and internal organs have been removed. Both sizes have a place in the market and both are produced. Some breeds are adapted to hot, humid climates and some tolerate the cold and snow better. A producer may like the color pattern or the docile nature of a particular breed and prefer to produce that breed.

Some breeds make excellent mothers, while other breeds grow rapidly and produce high-quality, meaty carcasses. Because of this, some breeds are referred to as sire breeds and some are referred to as dam breeds. A sire is the father and the dam is the mother.

Most animals produced for slaughter are crossbreeds. A crossbreed is an animal with parents of different breeds (Figure 14–4). A crossbreeding program helps producers to take advantage of the good points of both types of animals.

The beef industry has four major phases: purebred operations, cow-calf operations, stocker operations, and feedlot operations. Purebred cattle are produced in the first phase of the industry. The term purebred means that the animal is of a certain breed and has only that breed in its ancestry. Purebred

THE LIVESTOCK INDUSTRY

FIGURE 14-4
A crossbreed is an animal with parents of different breeds.

FIGURE 14-5
The purpose of a cow-calf operation is to produce calves to be grown out for market.

animals are eligible for registration, which certifies that the animal is purebred. The purpose is to produce what is known as seed-stock cattle. These cattle are used as the dams and sires of calves that will be grown out for market. These are usually those that are of the sire type or dam type of animal.

The second phase is the cow-calf operations, where the calves are produced. The calves are weaned from their mothers and sold to the next phase of the industry (Figure 14-5). Most of these calves are crossbred animals from purebred parents of different breeds. A large part of this industry is centered in the southern and western states. The mild winters of the

FIGURE 14-6
Stocker operations provide a step between weaning and the feedlot.

South are ideal for calving in the winter. In the West, producers can take advantage of the vast amounts of government lands that are open to grazing for a small fee. Often, cows are left on free range (not fenced in) to have their calves; the calves are then rounded up, weaned, and sold.

Calves are usually sold upon weaning. They are weaned at about six to seven months in age. They generally weigh approximately 500 to 600 pounds. At this age and weight they are ready to begin feeding for market.

The next phase of the industry is stocker operations, a step between the weaning of the calves and the finishing or fattening of the animals prior to slaughter. The stocker purchases the animals from the cow-calf producer and sells them to the feedlot operator. The stocker provides a transition period for the calves between the time they are weaned and before they are put in the feedlot (Figure 14–6). The calves are fed a ration, which is the amount of food necessary during a 24-hour period, that ensures that they make sufficient gains to be placed in the feedlot.

It is not uncommon for feedlot owners to also run stocker operations. Such arrangements are economical in that fewer transportation costs are

FIGURE 14-7

Cattle in a feedlot are fed a high-concentrate ration.

incurred if the two types of operations are close together. The trend in the industry has been away from the stocker industry as new research reveals production methods that allow cows to wean heavier calves.

The feedlot operation is the final phase before the animals are sent to slaughter. Here the animals are fed on a high-concentrate ration designed to put on the proper amount of fat cover (Figure 14-7). A concentrate is a feed that is high in grain content. Grain contains a lot of carbohydrates and helps an animal to fatten. A roughage is a feed such as grass or hay that is high in fiber. The producers usually want their animals to be marketed when the cattle reach a sufficient fat cover to allow the animals to grade "low choice."

Many feedlots are situated in the Midwest because this section of the country produces the most grain. It is usually more economical to feed the animals there rather than ship the grain across country. An exception is the state of Texas, which has more feedlots than any other state. When the animals are slaughtered, they are generally around 18 to 24 months in age and can weigh from 800 to 1,500 pounds. This age and size offer consumers the type of beef they prefer.

FIGURE 14-8
Pork production and consumption rank second only to beef in the United States.

The Pork Industry

As a producer and consumer of pork, the United States ranks behind Asia and Europe. Every year Americans produce over 85 million head of hogs. In terms of meat, pork production and consumption rank second only to beef in this country (Figure 14–8). Pork is distributed throughout the country, although some groups, such as Moslems and Jews, do not eat pork for religious reasons.

Much of the pork produced in this country comes from the midwestern states of Iowa, Illinois, Indiana, Minnesota, and Nebraska. These are the states that produce corn, the major grain fed to swine. Also there are large numbers of pigs in the South, where mild winters help lower the cost of production.

There are not as many breeds of hogs as there are breeds of cattle grown in this country. Breeds are categorized as mother or sire breeds. The mother breeds include Landrace and Yorkshire, which have a large number of pigs per litter and produce lots of milk for their young (Figure 14–9). The sire breeds, such as the Duroc and the Hampshire, characteristically grow rapidly and produce well-muscled, meaty carcasses.

Most of the market hogs produced are crosses of the various breeds. For example, a Yorkshire female mated to a Duroc male provides an excellent cross.

FIGURE 14-9

A Landrace female makes a good mother and is considered a dam breed.

FIGURE 14-10

Pigs are generally born in a farrowing crate to protect them from being crushed.

The Yorkshire sow provides a large litter and plenty of milk and the Duroc boar sires pigs that are vigorous and fast growing. In addition, crossbred pigs are healthier and do better than purebred pigs.

Two phases of the industry are the farrowing and finishing operations. The two phases can be separate, or they can be operated together. Some producers prefer to raise only feeder pigs (pigs that are sold shortly after weaning), and some prefer to buy feeder pigs and finish them as their only operation.

Most pigs are farrowed in climate-controlled houses where the mother is kept in a crate to prevent her from injuring the pigs as she lies down (Figure 14–10). Good producers make quite an effort to provide an environment that is clean, dry, and comfortable for both the mother and the piglets. The pigs are generally weaned from the mother at six weeks, although they may be as young as three weeks or as old as eight weeks.

After weaning, the pigs are placed together with pigs of similar age and size in what is called a confinement operation. This means that the hogs are

CAREER DEVELOPMENT EVENTS

If you have an interest in livestock and would like to work in the livestock industry, you could benefit from participating in several FFA career development events. The FFA Livestock Judging Event is for teams consisting of four members. Each member competes individually and the final team score is calculated by adding the three highest individual scores. The FFA Livestock Judging Event aids classroom instruction by teaching students practical skills in selecting quality livestock. All livestock producers know that the selection of quality livestock is essential to be competitive and profitable in the industry. Some skills and talents you will need to participate in the event are understanding the breed characteristics of beef, swine, and sheep. The ability to rank classes of animals and give reasons for placement of animals is also required. Team members independently evaluate and place each livestock class. After the judging, students are asked to give the reasons they used to place the livestock. Points are awarded for each class placing and correct reasons. This FFA event is usually held at the chapter, area or district, state, and national levels.

Another FFA team event related to the livestock industry is the Meats Evaluation Career Development Event. Like the Livestock Judging Event, teams are made up of four FFA members. As a participant in this event, you will gain valuable knowledge of meat identification and grading. This knowledge will benefit you in the future if seeking a job in the meatpacking industry or if you are a consumer of meat. Preparation for this event combines team practice and individual study. The team should practice as often as possible. In most career development events, students identify primal cuts of beef and pork; retail cuts of pork, beef, and lamb; and quality or yield grades of carcasses. By visiting meat processing plants and local grocery meat departments, the identification of wholesale and retail cuts can be reviewed.

kept in a pen together, rather than running loose on a pasture. Sufficient space is allowed for the pigs to be comfortable and to grow at a fast rate (Figure 14–11). The pigs should be finished (reach the proper market weight and condition) at about 20 weeks. Packers like to buy market hogs that weigh in the range of 220 to 260 pounds.

In the meats judging career development event, students test their knowledge of meat quality and yield grades as well as the identification of different cuts of meat. *Courtesy of Blane Marable, Georgia Agricultural Education.*

Each contestant will need to study for the written test relating to meat selection, storage, cookery, nutrition, and safety. Situational problems involving the least-cost formation of a batch of meat products such as hamburger or bologna are given. The student works through procedural questions and the actual determination of the least-cost price. The FFA Meats Evaluation Career Development Event develops employment knowledge with practical skill applications preparing students for entry-level employment in the meatpacking industry.

Students may compete in individual events in this area through the FFA Proficiency Awards. All FFA members who complete SAEs as part of their agriculture education instruction are eligible for the award. Livestock proficiency awards are available in beef production, dairy production, diversified livestock production, and sheep and swine production. Chapter winners may compete at the state level, where the winner earns a cash award and a plaque, plus travel expenses to the National FFA Convention. The top FFA member in the nation in each award area is recognized and awarded at the National Convention.

FIGURE 14–11
Hogs are grown to market weight on a finishing floor.

The Sheep Industry

Compared to beef and pork, Americans eat relatively little lamb and mutton. Lamb refers to meat from a sheep that is less than a year old; mutton refers to meat from a sheep that is over a year old. In many parts of the world, lamb and mutton are a basic part of the diet. The U.S. per capita consumption of mutton and lamb is only about 2.5 pounds (Figure 14–12). Of this, about 95 percent is lamb and only about 5 percent is mutton. Americans seem to have never developed a taste for the stronger-flavored mutton.

One advantage of producing lambs for market is that very good-quality lambs can be produced on grass without having to be fed a lot of expensive grain. Although an increasing number of lambs are being fed on grain in feedlots, roughages still make up about 90 percent of all the feed consumed by sheep.

The leading states in sheep production are Texas, California, Wyoming, Colorado, South Dakota, Montana, New Mexico, Utah, and Oregon. Sheep can make better use of lower-quality forage than can cattle. For this reason, they can be successfully grazed on poorer grazing lands of the desert areas

THE LIVESTOCK INDUSTRY

FIGURE 14-12
Annual consumption of lamb is only about 2.5 pounds per capita in the United States.

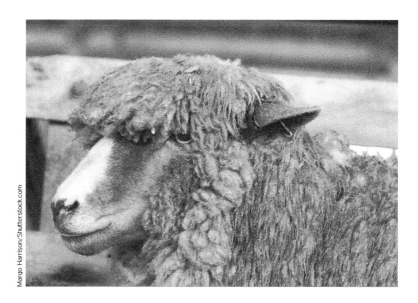

FIGURE 14-13
The Dorset sheep is an example of a medium wool breed of sheep.

of the West. In addition, a drier climate helps reduce parasite and disease problems associated with sheep that are grown in the more humid areas.

Sheep breeds are generally grouped according to the type of wool the animals grow. Wool type is broadly classified as fine wool, long wool, medium wool, hair, and fur. The medium wool breeds are used most often to produce lambs for slaughter. Medium wool breeds commonly used to produce slaughter lambs are Suffolk, Hampshire, Southdown, and Dorset (Figure 14–13).

The Horse Industry

Humans have used horses for transportation, work, and war as far back as recorded history goes. Almost all civilizations at one time relied on horses or donkeys. In this country, much of our history has been built around power supplied by horses and mules (Figure 14–14).

The number of horses and mules in this country grew until the 1920s, when the car, truck, and tractor caused a sharp decline in their numbers. From that time until around 1960, the numbers steadily declined. Since the 1960s, the number of both horses and mules has increased dramatically. While some horses and mules in the United States are used for work, most of the animals are used for recreational purposes.

In modern times horses are generally categorized as light horses, draft horses, or ponies. Light horses are animals that weigh 900 to 1,400 pounds. These horses are further divided according to use. Gaited saddle horses and walkers are used primarily for pleasure riding and show. They have different ways in which they walk or trot that seem graceful and flowing. Driving horses are used to pull carriages that are either four or two wheel and are used for recreation or for show.

FIGURE 14–14

Much of our history has been built with the power of horses and mules.

THE LIVESTOCK INDUSTRY

Stock horses are used all across the country to work cattle or other animals. Racehorses are used to compete with each other as a spectator sport.

Draft horses are those breeds that weigh over 1,400 pounds. At one time, these animals provided the power for pulling heavy loads such as wagons, plows, and other agricultural implements. Today, these animals are used in pulling competitions, shows, and parades (Figure 14–15).

Ponies are breeds of horses that weigh 500 to 900 pounds and are less than 14 hands high when mature (Figure 14–16). A hand is a measurement equal to 4 inches. While these are used to pull carriages and for show, the majority of ponies are used as horses for children.

Altogether the horse industry in this country is a $15 billion industry. Surprisingly, horse racing ranks third behind baseball and auto racing among spectator sports in the United States. Each year over 75 million people attend thoroughbred and harness racing. Each year there are around 7,000 horse shows in this country where young people and adults compete in a variety of different events. Many horses are owned as individual saddle horses that are used as recreation and are never raced or shown.

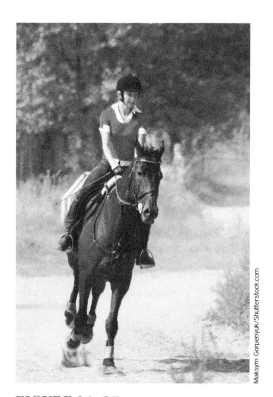

FIGURE 14–15
Today most horses are raised for recreation.

FIGURE 14–16
Ponies are breeds of horses that weigh between 500 and 900 pounds at maturity.

Small Animals

A growing industry in the United States is that of producing small animals. These animals range from rabbits raised for meat to animals such as gerbils and mice raised for pets. Animals such as guinea pigs and rats are raised for research. These animals are used to test everything from drugs to toothpaste.

Rabbit Production

People have raised rabbits for food for hundreds of years. Although most rabbits are produced by part-time growers, there are several large commercial operations in this country. Americans consume around 10 to 13 million pounds of rabbit meat per year. Some of the meat consumed in this country is imported from Europe, where the rabbit industry is larger. France is the leading producer.

There are several advantages to raising rabbits. Rabbits are easily raised under almost any climatic condition. Most are raised indoors in cages called hutches that consist of woven wire with boxes for the rabbits to sleep in and to have their young (Figure 14–17).

Many restaurants now offer several dishes that are prepared using rabbit meat. The USDA points out that rabbit meat is one of the most nutritious meats

FIGURE 14–17
Rabbits are raised in cages called hutches.

available. Not only is it high in protein and low in fat and cholesterol, but also it is easily digested and is very flavorful. In addition to the meat, rabbits are used for several purposes. The fur is used in making coats, hats, liners for boots, and for toys. Scientists use the animals in medical research and other areas. Many animals are sold as pets because of their docile nature, clean habits, and cuddly fur.

Rabbits can be a relatively inexpensive source of meat, but Americans do not consume as much rabbit meat as other peoples of the world. While we may accept the production and slaughter of cattle, hogs, and chickens, the thought of slaughtering cute, cuddly animals such as rabbits is repulsive to many.

SUMMARY

The livestock industry in this country is broad and diverse. One of the biggest reasons is that Americans eat a lot of meat every year. They also raise animals such as horses and ponies for recreation. Most experts say that in the future it will become even more diverse. In the future, production will expand to include some exotic animals (Figure 14–18). Animals such as deer and alligators may be grown for food. This may seem far-fetched, but remember that at one time cattle, horses, and sheep were wild animals before humans tamed them.

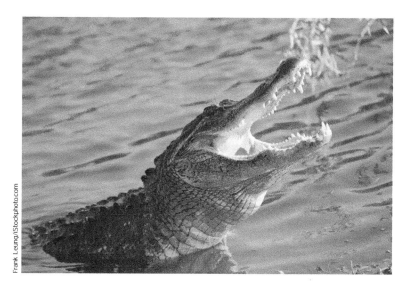

FIGURE 14–18

In the future, animal agriculture may expand to include animals that are now considered to be exotic.

CHAPTER REVIEW

✲ Student Learning Activities

1. Make a list of all the products in your home that come from animals. Determine what type of animal was used for the product. For example, a leather belt was probably from a beef animal and a wool sweater was probably from a sheep.
2. Report on the most widely grown agricultural animal in your state. List the reasons why this animal is popular in your state.
3. Design a plan for growing animals that are not grown in your area. Give your reasons for growing the animals, why your area would be suitable, and how you would market the animals.
4. Go to the Internet and research a breed of livestock. Determine where and when the breed originated. What are the characteristics of the breed?

✲ True/False

1. _____ The livestock industry in the United States is a tremendously large industry.
2. _____ Our country is not suited for growing livestock.
3. _____ Meat is among the most nutritionally complete foods that humans consume.
4. _____ The average size of the beef herds in the United States is about 50 animals.
5. _____ Most feedlots in this country are situated in the South.
6. _____ Ostriches, emus, deer and alligators may be grown for food.
7. _____ Calves are usually sold on weaning.
8. _____ The United States leads the world in the production and consumption of pork.
9. _____ Americans eat more mutton than lamb.

✲ Multiple Choice

1. The largest segment of the meat industry is
   a. beef
   b. poultry
   c. fish

2. About 90 percent of all feed consumed by sheep is
   a. mutton
   b. corn
   c. roughage

3. The major grain fed to swine is
   a. corn
   b. wheat
   c. roughage

4. Horses that weigh 900 to 1,400 pounds are called
   a. light
   b. purebred
   c. crossbred

5. The third largest spectator sport in the United States is
   a. horse racing
   b. football
   c. auto racing

6. Most rabbits are raised indoors in cages called
   a. hutches
   b. sires
   c. dams

7. A substance found in meat that can cause heart problems is called
   a. mutton
   b. roughage
   c. cholesterol

8. Over three-quarters of all of the money earned from the meat industry comes from
   a. chicken
   b. beef
   c. turkey

9. The mild winters of the South are ideal for
   a. calving
   b. racing
   c. wool production

10. The step between the weaning of the calves and the finishing, or fattening, of animals prior to slaughter is
    a. registration
    b. feedlot operations
    c. stocker operations

Discussion

1. List some uses for animals.
2. Why are health authorities expressing concern over the consumption of so much meat in the United States?
3. What vitamins do meat and animal products provide?
4. How do livestock producers choose the breed they grow?
5. What are the four major segments of the beef industry?
6. What is an advantage of producing lambs?
7. Why is it better to raise sheep in a drier climate?
8. How is wool classified?
9. Why did the number of horses and mules used in this country decline?
10. What are the advantages of raising rabbits?

CHAPTER 15
The Dairy Industry

Student Objectives

When you have finished studying this chapter, you should be able to:
- Explain the importance of the dairy industry.
- Discuss the controversy surrounding the use of BST.
- Tell how milk is produced.
- Explain how milk is processed and graded.
- Discuss the cheese-making process.
- Tell how yogurt is made.

Key Terms

dairy products
selective breeding
BST
fermentation
milkfat
homogenization
pasteurization
cheese
yogurt

FIGURE 15-1

Around 13 percent of all sales of farm commodities are from dairy products.

A large portion of agriculture is the dairy industry. This industry includes all of the products made from milk. Around 13 percent of all sales of farm commodities are from dairy products (Figure 15-1). Americans use a lot of dairy products. On average, we consume 28.8 gallons of milk, 26.8 pounds of cheese, 16 pounds of ice cream, 4.3 pounds of butter, and 4.3 pounds of yogurt per person, per year. These products are high in protein and contain essential minerals and vitamins. Milk has been described as nature's most perfect food. It is the only substance intended by nature for no other purpose than as food. Milk is obtained from cows that produce it to feed their calves (Figure 15-2). Because it is the only food a very young animal receives, it must be filled with nutrients. Humans make use of this nutritious food.

Selective breeding of dairy cattle has developed cows that produce much more milk each year than a calf can consume. Selective breeding means that producers have selected the best animals to breed in order to produce the most desirable offspring. This allows us to use the surplus for our food. There are about half the number of dairy cows in this country than there were in 1950. However, the amount of milk produced remains about the same. This means that modern dairy animals produce twice the milk their ancestors did.

FIGURE 15-2

Milk is intended by nature to be a food for young animals.

In recent years, consumers have been concerned over a substance given to dairy cattle to increase production. This substance, called bovine somatotropin or BST, is a naturally occurring hormone that helps to stimulate milk production. Scientists have known for many years that cows that get additional amounts of BST significantly increase their milk production. Until recently, this substance was scarce and very expensive. However, because of genetic engineering, the hormone can now be produced quickly and cheaply. The use of BST has brought about controversy over the safety of the substance. Some say that no studies have been conducted to determine the long-term effects of drinking milk from cows that have been given BST. They point to the fact that large doses can cause inflammation of the cows' udders and that this may be a sign that the hormone is not safe. Proponents point out that the cows are only given very small doses and that no ill effects have been discovered. Also, the National Institutes of Health and the Food and Drug Administration have declared BST to be safe both for cows and humans who consume the milk. All milk contains a small amount of BST. BST is a naturally occurring substance. Milk from cows given BST has only a minute amount more BST than milk from cows that have had no artificial BST. All BST in the milk is digested and none gets into the bloodstream of humans.

Milk Production

Most of the dairy cattle in the United States are Holstein (Figure 15–3). This breed produces more milk than most other breeds. Some breeds, like the Jersey, produce milk with a higher fat content. This was once a desirable characteristic. Today, however, consumers want foods with less fat content. Most of the milk produced in the United States comes from Wisconsin, California, New York, Minnesota, and Pennsylvania. These leading states have the climate

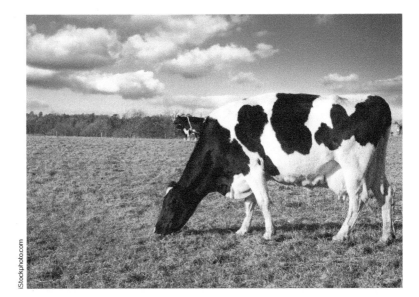

FIGURE 15-3

Most dairy animals in the United States are of the Holstein breed.

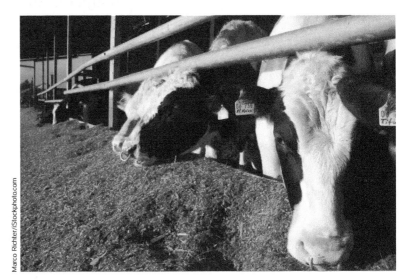

FIGURE 15-4

Dairy cows are fed on silage, which is made from chopped corn.

necessary for cows to do well and large cities where the milk is marketed. However, every state produces and processes some milk.

Most dairy animals are fed on pastures. Grass makes a relatively inexpensive feed for the cows. Some dairies now feed their cows in lots rather than pastures. These cows are fed on silage. This feed is made by chopping green corn (ears, stalks, and leaves) (Figure 15-4). The chopped corn goes through a process known as fermentation, which is a chemical change that helps to preserve the feed.

In order for cows to produce milk they must have a calf each year. Remember that the function of milk is to feed the young animal. The cycle of birth stimulates the mammary system of cows to produce milk. Most of the cows are bred artificially. This offers several advantages. By using artificial insemination (AI), producers do not have to go to the trouble and expense to keep a bull. Also, problems with sexually transmitted diseases are eliminated. AI gives the advantage of being able to select semen from a wide variety of superior bulls at a relatively low price. Daughters from these bulls can be used for replacement heifers and the quality of the herd can be raised.

When the calf is born, the first milk that is produced is different. This milk, called colostrum, is loaded with the mother's antibodies (Figure 15–5). These help the young animal to ward off disease during the first few days of life. Also, colostrum contains substances that help to open up the calf's digestive system and get it to function properly. Although this milk is essential to the young animal, it is considered unfit for human consumption. Cows that have just given birth are separated and the colostrum is not mixed with milk intended for the market.

The cows are milked in an area called a parlor. These areas are scientifically designed so the cows can be handled gently. They come into the parlor and enter a station where they obtain feed. The cow stands still while she is being milked. A milker enters the parlor and milks a small amount of milk

FIGURE 15–5
The first milk a newborn calf gets is called colostrum.

by hand into a container called a strip cup. This gives the milker an opportunity to check for problems in the milk. Lumps or blood in the milk could indicate a disease called mastitis. This disease is caused by injury to the cow's udder. Milk from cows with mastitis is not used for human consumption. Milking and discarding the first two or three squirts of milk also helps to cleanse the teat.

The udder is then washed using a warm water solution and dried. The teat cups are then attached and the milking begins. These cups are lined with a soft material that is attached to a tube that gently pulls a vacuum on the cup to draw milk from the teat (Figure 15-6). The teat cup fits snugly on the cow's teat and pulsates by means of a vacuum on the lining of the cup. The milk is all removed in three to six minutes, depending on the individual cow and the amount of milk she gives. Care is given that the cups are left on the proper amount of time. Too little time and the udder will not be milked out, and too much time and injury to the udder can result. The teats are then treated with a disinfectant and the cows are released. The teat cups are cleaned between cows to prevent the spread of disease.

The milk is drawn through lines and into a holding tank. In modern dairies the cooling of the milk begins in the lines. The lines are refrigerated so that the milk is cooled as it is transported to the tank. Here it is rapidly cooled to around 40 degrees F to prevent the multiplication of bacteria and to prevent the milk from spoiling. After all the cows have been milked, the lines, teat cups, and all of the equipment are thoroughly cleaned. About every other day, the milk is picked up by a tanker truck and hauled to the processing plant. At the plant, the milk is tested for the number of bacteria, drug residue, and any indications of infection. Milk with high levels of bacteria, drug residues, or signs of infection is discarded.

When the milk arrives at the processing plant it is thoroughly filtered to remove any foreign particles. The milk is allowed to sit so the cream may be

FIGURE 15-6

A milker gently attaches teat cups that draw out the milk.

FIGURE 15-7
Milk is sold as whole, lowfat, and skim milk.

FIGURE 15-8
In the process of pasteurization, harmful bacteria are destroyed. The large tanks heat the milk to the proper temperature.

removed from milk to be sold as lowfat milk. Cream is the part of the milk that contains fat. As consumers are becoming more conscious of the amount of fat in their diet, they want milk that is lower in milkfat than whole milk. In recent years, sales of lowfat and skim milk have increased sharply (Figure 15-7). Lowfat milk is milk that has had the percentage of milkfat lowered to between 0.5 and 2 percent. Skim, or nonfat, milk is a milk containing less than 0.5 percent milkfat. The milkfat that is removed from the milk is then used to make ice cream and other cream products.

Whole milk contains around 3.5 to 4% milkfat. The globules of fat are what makes the cream float to the top of raw, unprocessed milk. In a process called homogenization, the large cream globules are reduced in size to the size of the milk globules. The processed milk, called homogenized milk, will not separate out.

To kill any harmful organisms in the milk, the milk is first heated and then cooled in a process called pasteurization (Figure 15-8). In pasteurization, the temperature of the milk is raised to 149 degrees F for 30 minutes and then is promptly cooled to 39 degrees F. This kills harmful bacteria that might cause illness in humans. The time and temperature must be precisely controlled to protect the nutritive value and flavor of the milk.

Milk is graded according to the dairy from which it came. Dairies that sell Grade A milk must pass rigid standards that include cleanliness and other conditions. Only Grade A milk can be used as fluid or beverage milk. Grade B milk can only be used for manufactured dairy products. As the production of Grade A milk far exceeds the demand for fluid milk, Grade A is used in processing also. For pricing purposes, the milk is classified as Class I, II, or III. Class I is used for beverage consumption; Class II is used for manufacturing soft products such as ice cream, yogurt, and cottage cheese; Class III is used with Grade B milk in the processing of cheese, butter, and nonfat dry milk.

Cheese

The United States produces more cheese any other country in the world. There are literally hundreds of types of cheese. They are broadly classified according to the type of milk used and the type of processing. Cheeses are often named for the town or region where they originated. Most of the cheese produced in the United States uses cow's milk. In some parts of the world, goat's milk (called chèvre cheese) or

CAREER DEVELOPMENT EVENTS

The FFA offers many exciting and challenging experiences. Members who participate in team events spend many hours learning the basics in the classroom and practicing skills after school. The Dairy Cattle Judging and Dairy Foods competitions are two events that require cooperative efforts of team members. If you have an interest in dairy products, plan to own dairy cattle, have a Supervised Agricultural Experience (SAE) in the dairy industry, or work on a dairy farm, you would be a good team member for these events.

The Dairy Cattle Judging Career Development Event trains students in the selection of a quality dairy herd. Practice and experience are the only ways to develop the skills needed to select quality animals. A well-prepared team should understand what a quality dairy animal looks like and be familiar with dairy herd improvement records. Most Dairy Judging teams consist of four FFA members. Usually, through a competitive process, the agriculture teacher selects a team to represent the FFA chapter in competition. The FFA Dairy Judging event requires each team member to know the breeds of dairy cattle and traits that yield high milk production. During the event, each member is responsible for evaluating and placing groups of animals in numerical order. The student must also give reasons for each animal's placement. After each team member has evaluated and placed all dairy groups, individual scores are calculated. The top three scores are added to determine the total team score and placement. All winning teams from each state advance to the National Dairy Cattle Judging Career Development Event.

sheep's milk (called Roquefort) may be used in the process. Other cultures may make cheese from the milk of horses or camels.

Cheese contains one of the highest concentrations of nutrients of any food we eat. Only 3.5 ounces supplies about 36 percent of the protein, 80 percent of the calcium, and 34 percent of the fat recommended in our daily diet. It is rich in essential amino acids, minerals, and vitamins, and has a high caloric content.

Most cheese begins with milk that has been pasteurized to help prevent the multiplication of harmful bacteria. The milk is placed in a large vat

Students competing in the Dairy Foods Career Development Event must distinguish between grades of dairy products.
Courtesy of Carol Duval, National FFA Organization.

Another FFA event that helps prepare students for careers in the dairy industry is the Dairy Foods Event. Hundreds of products are manufactured by the dairy industry. Every dairy product is graded by its quality. It is important that workers in the dairy industry and consumers know what makes a high-quality product.

To participate in this event, a student must have a keen sense of sight, smell, and taste to distinguish grades of dairy products. Classroom study of milk content and flavor, classification of cheeses, and the parts and operation of milking equipment are essential. The skills required to rank dairy products in order of quality can only be developed by practice. The Dairy Foods Event is a great opportunity to prepare FFA members with skills required to be competitive dairy workers and knowledgeable consumers.

FIGURE 15-9
To begin the cheese-making process, milk is placed in a large vat and starter cultures of bacteria are added.

and a bacteria culture is added (Figure 15-9). This culture is called a starter culture because it starts the process of fermentation, which changes sugars to acids. These acids cause the proteins in the milk to coagulate, that is, to form into a solid. To further the process, an enzyme called rennet is added. An enzyme is a substance that speeds up or stimulates a chemical process. Rennet is obtained from the stomachs of calves. During this step, large paddles turn the milk to ensure that the starter bacteria and the rennet are evenly distributed.

The solid resulting from this step is called curd. The liquid is drained off and is called whey (Figure 15-10). Whey is used in processing human

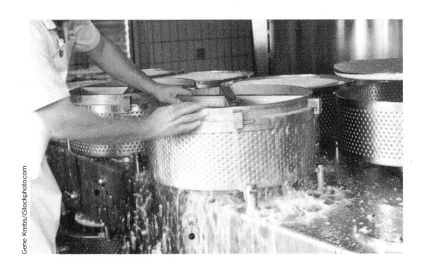

FIGURE 15-10
The solid is called curd and the liquid that is drained away is called whey.

food and manufacturing animal feed. The curd is cut into small cubes to increase the surface area to allow the drainage of the whey. After the whey is drained off, the curd sits until it forms a solid mass. The curd is then heated to cause the curd to contract and further expel the whey. The amount of heat and length of time the cheese is heated depend on the type of cheese being made. The cheese is salted and pressed into a metal form or a cloth bag.

The final step in cheese making is the curing or ripening of the cheese. The cheese is placed in an environment that is controlled for temperature and humidity. The specific conditions vary with the type of cheese being made. During this time, enzymes produced from the starter bacteria change the flavor, texture, and appearance of the cheese. The cheese is packaged in a coating of paraffin or is wrapped in cloth or plastic. It is allowed to age a specific amount of time before it is sold (Figure 15–11).

FIGURE 15–11
Cheese is coated, packaged, and stored for aging.

Yogurt

Yogurt is also processed from milk. This food product has been around for hundreds of years. It has gained in popularity during the past few years. When frozen, it is delicious and is lower in calories than ice cream. Many people prefer frozen yogurt to ice cream. Yogurt comes in a wide variety of flavors and is often flavored with fruit.

Processing begins by heating concentrated milk. It is cooled rapidly and cultures of bacteria are added. These cultures cause the milk to ferment. In this process, the required acidity and the yogurt flavor are produced. The acidity helps to give the yogurt a tart flavor and also helps to preserve it. Before it is packaged, fruit or other flavorings may be added.

SUMMARY

Dairy is a very significant part of our agricultural industry. It provides many jobs in the production, processing, and distribution of milk and milk products. It also provides Americans with wholesome, nutritious, and good-tasting food. In the future, new uses may be found for milk products that will make the industry even more important.

CHAPTER 15 REVIEW

Student Learning Activities

1. Visit a large grocery store and make a list of all the different types of dairy products that are available. Interview the manager to find out which sell the best.
2. Bring in samples of whole, lowfat, and skim milk. Blindfold class members and see if they can identify each by taste. Ask them how each tastes and record their responses.
3. Choose a particular type of cheese and write an essay on the cheese. Include how the cheese is different from other cheese, where and how it originated, where it is made, and the process used in the manufacture.
4. Collect all the information you can about the use of BST with dairy cattle. Decide whether you believe the product should be used. Defend your position to the class.

True/False

1. _____ Around 50 percent of all sales of farm commodities are from dairy products.
2. _____ Milk has been described as nature's most perfect food.
3. _____ There are about half the number of dairy cows in this country than there were in 1950.
4. _____ BST is a hormone that is very expensive to produce.
5. _____ All milk contains a small amount of BST.

6. _____ Only a few states process and produce milk.

7. _____ In order for cows to produce milk they must have a calf each year.

8. _____ Colostrum is not used for human consumption.

9. _____ The United States produces more cheese than any other country in the world.

10. _____ Milk is the only substance that is intended by nature for no other use than as food.

✵ Multiple Choice

1. A very large portion of animal agriculture is the
    a. dairy industry
    b. pesticide industry
    c. wheat industry

2. The dairy industry includes all of the products made from
    a. cows
    b. colostrum
    c. milk

3. Dairy products are high in
    a. hormones
    b. mastitis
    c. protein

4. Most of the dairy cattle in the United States are
    a. Holstein
    b. Angus
    c. Sires

5. A relatively inexpensive feed for cows is
    a. wheat
    b. grass
    c. grain

6. Most of the cheese produced in the United States comes from
    a. goats
    b. cows
    c. artificial dairies

7. The enzyme rennet is obtained from calves'
    a. udders
    b. kidneys
    c. stomachs

9. Colostrum is loaded with
   a. chemicals
   b. antibodies
   c. acids

10. Mastitis is caused by injury to the cow's
    a. stomach
    b. udder
    c. kidneys

11. Whole milk contains about 4 percent
    a. colostrum
    b. calcium
    c. milkfat

Discussion

1. What are the advantages of artificial breeding?
2. Why is colostrum good for newborn calves?
3. How is milk classified?
4. What animals are used for their milk?
5. What nutrients are supplied by cheese?
6. What purpose does acidity have in yogurt?
7. Why is the dairy industry so important?
8. What does fermentation do to milk?
9. Why are bacteria added to milk to make yogurt?
10. Why is there concern about the hormone BST?

CHAPTER 16

The Poultry Industry

Student Objectives

When you have finished studying this chapter, you should be able to:
- Explain the importance of poultry to humans.
- List the different segments of the poultry industry.
- Explain how eggs are hatched commercially.
- Describe how broilers are produced.
- Explain how eggs are produced.
- Discuss the growing of turkeys and other poultry.

Key Terms

game birds
broilers
nutrition
incubator
brooder
layers

Humans have grown poultry for almost as long as they have been involved with agriculture. The earliest archeological evidence reveals that poultry was a part of the human diet. All through the history of the world's great civilizations is evidence that people ate chickens and other birds.

Modern Americans also eat a lot of poultry. Each year, the average person in this country consumes over 90 pounds of chicken and that amount has increased each year for the last 35 years. The amount of poultry meat consumed is rising each year. The reason is that poultry is considered to be healthier than some other meats. The amount of cholesterol and fat is lower in a pound of chicken or turkey than in a pound of red meat (Figure 16-1).

Americans also eat a lot of eggs. Each year we consume about 250 eggs per person. According to the American Egg Board, producers care for about 280 million laying hens and each hen produces about 250 to 300 eggs per year.

The poultry industry includes the raising of chickens, turkeys, ducks, geese, game birds (birds hunted for food), and squab (pigeons). By far, the largest segment of the industry is the growing of chickens. This segment is divided into two parts—broiler production and egg production.

FIGURE 16-1

Poultry is lower in calories and cholesterol than beef or pork.

Meat- and Egg-Producing Chicken Breeds

Meat

Egg

FIGURE 16-2
Broiler chickens are a different type than layer chickens.

At one time most farms raised chickens for both meat and eggs. Females were kept to lay eggs and the males were butchered for meat. Today the industry is more specialized. Two different types of birds are produced. Through selective breeding, chickens have been developed that are thickly muscled and produce more meat. Layers have been developed that produce eggs daily (Figure 16-2). They are more slender than the birds raised for meat. Layers convert more of their energy into egg production than do meat-type birds.

The Broiler Industry

Broilers are chickens that are produced for meat. They are generally marketed at about six to eight weeks of age and weigh 3 to 5 pounds. Since 1950, the production of broilers has increased dramatically. Now we can grow a broiler that is heavier on half the feed in half the time (Figure 16-3). This is mostly because of research in **nutrition** and management, and also because of selective breeding. This means a real savings for the consumer. The efficiency with which these animals can be grown means that they can be sold cheaper.

Most of the nation's broilers are grown in the southern states. The leading states in broiler production are Arkansas, Georgia, and Alabama. In these

FIGURE 16-3
We can grow a heavier broiler in half the time on half the feed than we could in 1950.

states the winters are mild and the chickens can be grown at a cheaper cost. Most broilers are raised on contract. This means that the producer supplies the house, utilities, and labor. A company supplies the producer with chicks, feed, medication, and other supplies. The company pays the producer for the broilers at the end of the grow-out period. This substantially reduces the risk for the producer.

Hatching the Chicks

The first step in the production of broilers is the hatching of chicks. Breeder birds are carefully selected for rapid growth, good feed efficiency, and heavy muscling. Almost all of the broilers produced are white (Figure 16–4). This is because birds that are dark in color have spots of color or pigmentation left where the feathers are removed after slaughter. These spots do not lower the quality of the meat in any way, but consumers are reluctant to buy chicken with spots on the skin.

Eggs are hatched in large commercial hatcheries. Eggs that are to be hatched have to be fertile. This means that the hens have to be bred by roosters or they are artificially inseminated. Hens that lay eggs for hatching are housed in facilities with laying boxes (Figure 16–5). These boxes are designed to keep the eggs clean. It is essential that hatching eggs be clean.

FIGURE 16–4

Almost all of the broilers produced are white.

FIGURE 16-5
This is a modern facility for hens that lay hatching eggs. Note the laying boxes in the background.

FIGURE 16-6
Hatching eggs must be kept clean and dry. These eggs are going into the incubator.

Even a small speck of foreign material on the shell can contain millions of microorganisms that can cause problems. Eggs that are dirty are not used for hatching. Dirt or any foreign material that must be scrubbed from the eggs usually renders the eggs unfit for hatching. The washing or scrubbing of the eggs removes the protective coating from the eggs and presses the dirt into the pores of the eggs. Hatching eggs are not allowed to become wet. Moisture on the eggs can also encourage the growth of bacteria (Figure 16-6).

Before they leave the farm where they are produced, hatching eggs are sorted to remove dirty, undersized, oversized, misshapen, cracked, or defective eggs. They are then fumigated to kill harmful organisms on the surface of the eggs. This process is precisely regulated to prevent harm to the eggs. The eggs are never allowed to become chilled and are stored at 70 to 80 degrees F until they are placed in the hatchery.

The eggs are transported to the hatchery on special racks that protect the eggs. At the hatchery, the eggs are removed from the carts and placed in the incubator. In nature, hens set on the eggs for three weeks to keep the eggs warm as the develop into baby chicks. Incubators are devices that replace the hen by keeping the eggs at an even temperature until the eggs hatch. Here the temperature is gradually increased to prevent the eggs from sweating. Sweating is the condensing of water vapor on the surface of the eggs and occurs when the temperature of the eggs is raised too rapidly. If moisture collects on the surface of the egg, bacteria may begin to grow and thrive.

In commercial hatcheries, the eggs are incubated in two separate rooms, the setting room and the hatching room. The eggs are placed in the setting-room incubator and closely monitored for temperature and relative humidity (Figure 16–7). The eggs are turned every day to ensure that a high percentage of the eggs hatch. The eggs remain in the setting-room incubator until one to two days before they are ready to hatch. They are then placed in the hatching room. The temperature is lowered slightly and the eggs hatch in chick holding trays.

Growing the Broilers

After the chicks are hatched, they are cleaned and dried. They are then placed in a warm environment. When they are about a day old, they are vaccinated and their beaks are trimmed. In this process, the tip of the beak is removed. This helps to prevent the chicks

FIGURE 16–7
Eggs are placed in an incubator and are carefully monitored for temperature and relative humidity.

FIGURE 16-8
Note the shortened beak on this chick.

FIGURE 16-9
Chicks are kept under a brooder to keep them warm.

from pecking each other (Figure 16–8). It does not harm the chicks and prevents problems throughout the lives of the chickens.

The chicks are placed in cardboard or plastic ventilated boxes and shipped to the producers. The chicks are placed in houses that hold from 10,000 to 50,000 or more chickens, depending on the size of the house. The houses are well insulated and well lit. Lights are kept on around the clock. This helps prevent a problem called cannibalism. For some reason, the chickens may begin to peck certain birds. This causes injury and often death. By trimming the beaks and leaving the light on, cannibalism can be lowered. The floor is covered with an absorbent litter consisting of shavings, sawdust, or other material.

At the start the baby chicks are placed under a brooder. Brooders are large pan-shaped heaters used to keep the chicks warm during the first days in the house (Figure 16–9). Brooders are usually suspended from the ceiling and can be raised or lowered depending on the temperature and the size of the chicks.

FIGURE 16-10
The chickens are fed through automatic feeders and waterers.

CAREER DEVELOPMENT EVENTS

The FFA offers agriculture students opportunities to apply classroom lessons and laboratory skills through competition. Matthew joined the FFA his first year in high school. One of his classes in agriculture was poultry science. Matthew's experience and classroom knowledge prepared him to be a winner in several FFA competitions. Matthew tried out for the Poultry Judging Career Development Event. FFA Poultry Judging is a team event. Matthew made the team along with three other FFA members. The team members spent many hours preparing for the area competition. After school, they practiced judging and placing classes of live birds, dressed poultry, interior and exterior of eggs, and processed chicken parts. At home they each studied assigned topics in the poultry science manual that their agriculture teacher would drill them on the following day. Matthew practiced public-speaking skills. After weeks of practice and study, the team competed in the area FFA Poultry Judging Event. Each of the four members had to judge and give reasons for classes of production hens, pullets, ready-to-cook broilers, chicken patties, interior eggs, and exterior eggs. They also had to identify 10 broiler parts and place four cartons of eggs, with an explanation of how they placed them. At the end of the event, each FFA member had to answer a 30-question test relating to management practices used in the poultry industry. After all teams completed each section of the event, individual scores were calculated. The top three individual scores for each team were added to determine the team score and placement. Matthew's team placed second and had the opportunity to advance to the state

Water is supplied by suspended waterers that the chicks quickly learn to use. They peck the nipple on the bottom of the waterer and obtain as much water as they need. Feed is given to the baby chicks by hand when they are small but is fed through automatic feeders as the birds get older (Figure 16–10). The feed is brought to the birds by means of a conveyer chain in the bottom of the trough. Every day, the equipment and birds are checked to ensure that the equipment is functioning properly and that the birds are doing well.

The birds are generally kept in the broiler house from seven to eight weeks. At this time, they weigh around 4.5 pounds and are ready for market. They

In the Poultry Judging Career Development Event, FFA members must determine the quality of various poultry products. *Courtesy of Blane Marable/Georgia Agricultural Education.*

event. Matthew's interest in the processing and marketing of poultry, development of a supervised poultry project, and devotion of time and energy has made him a winner and prepared him for a future career.

The Poultry Production Proficiency is one of many agricultural proficiency awards presented at the FFA's annual banquet. Applicants applying for the Poultry Production Proficiency Award must be involved in activities and enterprises requiring the application of classroom lessons to the profitable management of a poultry operation. All FFA members who had an SAE in the area of poultry competed for this award. Maintenance of a complete set of records, including income and expenses, inventory of poultry and related equipment, and knowledge of the poultry industry helps give students the winning edge.

are usually caught at night when the birds are less active. They are put into coops and loaded onto a truck for transportation to the processing plant. The producer then begins to clean the house to get ready for the next batch of chicks. Because the manure in the litter has a very high concentration of nitrogen and other elements necessary for plant growth, the litter is a valuable source of fertilizer.

Processing the Broilers

At the processing plant, the birds are inspected while they are alive. If they show any signs of ill health, they are condemned. This means that they cannot be used for human consumption.

After inspection, the chickens are slaughtered and the feathers are removed. The intestines and other internal organs are removed. The carcasses are then washed and inspected. Both the carcass and the internal organs are inspected for signs of disease or any other problems (Figure 16–11). If any problems are found, the carcass is condemned and the meat is not used for human consumption.

After the birds are dressed, they are prepared for sale. They may be packaged whole or they may be cut up into several pieces. The trend is toward selling different parts of the carcass. For example, a package may contain only breasts, legs and thighs, or wings. Some of the broilers may be used for

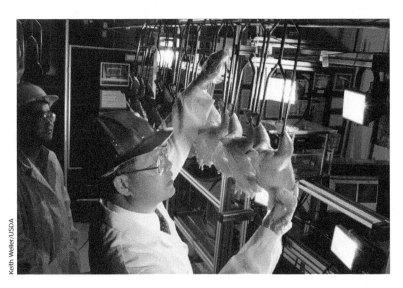

FIGURE 16-11
Carcasses are inspected for any signs of disease or other problems.

processed meat such as chicken franks, sausage, or bologna. Others may end up as chicken potpies or as prepared frozen dinners.

The Layer Industry

The other large segment of the poultry industry is the production of eggs. As mentioned earlier, hens, or layers, which produce many more eggs than their ancestors, have been developed through selective breeding. Today, the average layer produces over 250 eggs per year. Some layers produce brown eggs and some produce white eggs, but the vast majority of the eggs sold in the United States are white. Consumers simply prefer to buy white eggs. Most of the layers in this country are housed in cages within a large facility. The cages hold from 2 to 12 birds.

The hens are fed by means of an automated conveyer that carries the feed directly in front of the hens (Figure 16–12). Water is supplied by means of a narrow free-running trough or a nipple waterer that the hens learn to peck. As the eggs are laid, they roll onto a conveyer that periodically moves the eggs to a collection point where a worker gathers them and places them in flats (Figure 16–13). Dirty eggs are separated out and the eggs are refrigerated.

FIGURE 16–12

Layers are housed in cages where they are fed and watered by automation.

FIGURE 16-13
Eggs are brought from the laying cages on a conveyor. They are then sorted and packaged by workers.

At the processing and packing plant, the eggs are coated with a thin coat of light mineral oil. This coating prevents carbon dioxide from escaping from within the egg. This helps prevent the eggs from spoiling before they are consumed. The eggs are checked for cracks and interior spots in a process called candling (Figure 16–14). The eggs are passed over an intense light in a dark room and any blood spots or cracks in the shell will show up in the light. The eggs are then graded according to shape and size. Eggs of similar size are packaged together. The eggs are packaged and sent to the retail market or are sent to a processing plant where they are broken and processed.

Turkey Production

Turkey production is on the increase in this country. This is because of the many new ways of using turkey meat. Luncheon meat, sausages, wieners, and products resembling ham are made from turkey. Turkey meat is less expensive, and has fewer calories and cholesterol than many other meats. It also tastes good and is nutritious.

Turkeys are the descendants of the wild turkeys that were native to the Americas. Modern turkeys are very different from the wild variety. Like chickens, most domestic turkeys are white. Consumers do not want the coloring in the skin that comes with the bronze-colored turkeys. Also, turkeys raised for meat

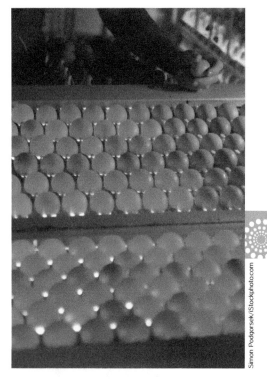

FIGURE 16-14
In a process called candling the eggs are checked for blood spots or cracks.

FIGURE 16-15
Turkeys may be grown on the range or in a house.

are much more muscular. The modern broad-breasted turkey is much more efficient at producing meat than its wild ancestors.

The majority of the turkeys in the United States are grown in the western part of the North Central region, the South Atlantic region, and the Pacific region. Unlike chickens, turkeys seem better able to tolerate cold weather than hot weather. Most are produced by small operations of 30,000 birds or fewer.

There are two major ways of growing turkeys—in confinement houses and on ranges (Figure 16–15). Confinement houses offer the advantages of environmental control of temperature and humidity. The open range offers the advantage of being less expensive. Turkeys on range can stay outdoors completely or they are provided with housing where the birds can get shelter when they want. Producers usually move the range every three to four years to help prevent problems with disease and parasites.

Other Poultry

Although chickens and turkeys make up by far the largest portion of the poultry industry, other types of poultry are also grown. In some parts of the world, poultry such as ducks and geese make up a major portion of the total poultry output. In China and Southeast Asia, and [parts of] Europe ducks and geese make up a large part of the overall diet (Figure 16–16).

In some ways, these birds are actually easier to raise than chickens and turkeys. They withstand harsh weather better and resist disease better than

FIGURE 16-16
In some parts of the world, ducks and geese make up a major part of the overall diet. These geese are on a farm in Denmark.

FIGURE 16-17
Pheasants and quail are raised for use in restaurants. These young quail are being raised for that purpose.

turkeys and chickens. Another advantage of growing geese and ducks is that the feathers are used to make bedding and other goods. In this country, most ducks and geese are raised in small flocks by hobbyists or part-time producers. Most of the meat sold goes to the restaurant trade or gourmet food market.

About the only other poultry production of any significance is that of growing quail and pheasant. Both of these birds are grown for the restaurant trade and the gourmet food market (Figure 16–17). Also, they are raised for restocking wildlife areas. Each year thousands of quail and pheasants are released in the wild to provide birds for hunting. This helps replenish the areas with game that is difficult to produce in the wild.

SUMMARY

As the world's population increases, the growing of poultry will become more important. Poultry is efficient to raise and people in most cultures eat poultry. Many developing countries will depend on poultry to help in the development of their agriculture. In America, we will continue to eat more poultry and poultry products.

16 CHAPTER REVIEW

✸ Student Learning Activities

1. Make an inventory of all the foods in your house that contain eggs. Do not forget foods such as pastries and mayonnaise. Share your list with the class.
2. Over a period of a week, keep track of all the meat products eaten by your family. Calculate the percentage of meals that contain poultry products. Include processed meats such as luncheon meats and hot dogs that may contain chicken or turkey.
3. Research the possibilities of raising exotic birds such as emus and ostriches. Provide a rationale for growing the birds and the advantages and disadvantages. Report to the class.
4. Go to the Internet and find the latest information on eating eggs. What conclusions can you draw?

✸ True/False

1. _____ Humans have been growing poultry for almost as long as they have been involved with agriculture.
2. _____ Each year the average person in the United States consumes over 90 pounds of chicken and 22 dozen eggs.
3. _____ At one time, most farmers raised chickens for both meat and eggs.
4. _____ Most of the nation's broilers are grown in the northern states.
5. _____ Almost all of the broilers produced are tan.
6. _____ The vast majority of the eggs sold in the United States are white eggs.

7. _____ Turkeys cannot tolerate cold weather as well as chickens.

8. _____ In China and Southeast Asia, ducks and geese make up a large part of the overall diet.

9. _____ Chickens make up the largest part of the poultry industry. The second most widely produced birds are geese.

10. _____ Each year thousands of quail and pheasants are released into the wild to provide birds for hunting.

Multiple Choice

1. The meat that is considered to be healthier than most meats is
    a. beef
    b. pork
    c. poultry

2. The largest segment of the poultry industry is raising
    a. turkeys
    b. fish
    c. chickens

3. Raising chickens is divided into two parts, broiler production and
    a. egg production
    b. candling
    c. beak trimming

4. Chickens that produce eggs daily are called
    a. layers
    b. brooders
    c. hatchers

5. Chickens produced for meat are called
    a. layers
    b. hatchers
    c. broilers

6. One of the first steps in the management of broilers is
    a. candling
    b. beak trimming
    c. hatching

7. Eggs that are to be hatched must be
    a. fertile
    b. thoroughly washed
    c. fumigated

8. At processing and packaging, eggs are lightly coated with
   a. cooking oil
   b. olive oil
   c. mineral oil

9. Candling is a process that checks for defects in
   a. feather pillows
   b. eggs
   c. fish

10. Areas designed to keep eggs clean are
    a. laying boxes
    b. incubators
    c. conveyers

Discussion

1. How are breeder birds selected?
2. Why are almost all of the broilers white?
3. Why is it essential that hatching eggs be clean?
4. What are the two rooms in a commercial hatchery?
5. Why are lights kept on in a chicken house?
6. Why is the litter in a chicken house a valuable fertilizer?
7. Why are eggs coated with mineral oil?
8. What are the two major ways of growing turkeys?
9. What animals are raised in the poultry industry?
10. What is meant by raising broilers on contract?

CHAPTER 17
The Science of Aquaculture

Student Objectives

When you have finished studying this chapter, you should be able to:
- Discuss why aquaculture is a growing industry.
- Explain why fish are efficient agricultural animals.
- Explain the steps involved in the production of both catfish and cool-water fish.
- Discuss the production of crustaceans for human consumption.
- Describe the nonfood segments of aquaculture.

Key Terms

aquaculture
ectothermic
warm-water fish
brood ponds
roe
fry
fingerlings
dissolved oxygen
cool-water fish
crustaceans
ornamental fish

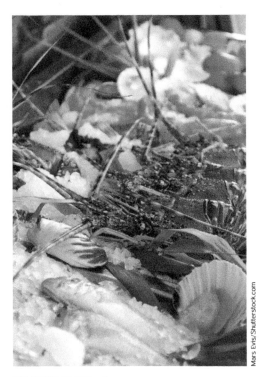

FIGURE 17-1
The amount of fish and seafood eaten by Americans is increasing each year.

One of the newest and most rapidly developing parts of agriculture is that of growing animals in water. This science is called aquaculture. Aqua means water and culture means growing or caring for. Many of the foods we eat are grown in water. You probably ate some type of fish this past week. Perhaps you enjoy seafood such as oysters or shrimp. The amount of fish and seafood eaten by Americans is increasing each year (Figure 17-1). Each year the average consumption per person in the United States is around 16 pounds of fish and seafood. Experts tell us that eating fish is nutritious and makes a good addition to the diet. It is high in protein and low in calories. The fat content is lower than that of red meat, and this helps prevent health problems. As people become more health conscious, the demand for fish and seafood will increase.

People have grown fish for food for thousands of years. The ancient Chinese raised fish, as did the Egyptians and later the Romans. This provided these people with a ready supply of fish. Those of you who enjoy fishing know that on certain days, even the best angler cannot catch any fish. If the fish have been raised in small ponds, they can be caught at any time.

Only during the past 30 years has aquaculture become a major industry in this country. Up until that time, most of the fish sold in the markets were caught from the wild (Figure 17-2). Large seagoing trawlers harvested the fish from the oceans. This supplied the world with a relatively inexpensive source of fish. However, as the world population increased and the demand for fish increased, more and more fish were caught from the ocean. A few years ago, the maximum amount of fish that could be taken from the oceans in a year was reached. The oceans simply cannot sustain enough fish to meet the demand. This created a greater demand for fish than there was a supply of fish. As a result the price of fish went up. To provide people with a supply of fish, fish farms came into production.

FIGURE 17-2
Up until a few years ago, most of the fish we ate came from the ocean. These fish have just been harvested from the ocean.

At first, not much was known about growing fish for food. But scientists at several of our universities conducted extensive research in how to produce fish. This resulted in an industry that is now providing us with an inexpensive supply of fish.

Fish are very efficient farm animals. This means that a pound of weight can be put on these animals with less feed than other agricultural animals. Fish can gain a pound of body weight on less than 2 pounds of feed. A steer may require 8 or 9 pounds of feed to gain a pound of body weight. This is partly because fish are ectothermic. This means that they depend on their environment for their body temperature. In other words, the water gives them their body temperature. Most other agricultural animals, such as chickens, pigs, and cattle, are endothermic and rely on heat generated from their feed to maintain their body temperature. Land animals also have to use a lot of energy moving about. Aquatic animals are suspended in the water and use less energy in moving about (Figure 17-3). All this saved energy can be put into growing and gaining weight.

Catfish Production

By far the most important food fish grown in this country is the catfish. In 1960, there were only 600 acres of commercial catfish ponds in the

FIGURE 17-3
Aquatic animals are suspended in water and use less energy moving about.

FIGURE 17-4
Catfish has replaced cotton as the number one crop of the Mississippi Delta region.

United States. Today, there are over 150,000 acres of catfish ponds. These fish are raised in the southern part of the country because they are warm-water fish. In fact, catfish grow best when the water temperature is around 85 degrees F. Mississippi leads the nation in catfish production. For many years, the fertile Mississippi Delta region produced cotton as the major crop. Today, catfish has replaced cotton as the number one cash-producing agricultural enterprise (Figure 17-4).

Catfish are different from most other fish. They have a smooth skin instead of scales. They also have long whiskers (known as barbels) extending from their mouths. They are meaty and have a mild

THE SCIENCE OF AQUACULTURE

flavor. They can be cooked in a variety of ways ranging from deep-fat frying to broiling. Also, catfish fit well into farming operations because they are hardy and tolerate relatively low oxygen content in the water (Figure 17-5).

The production of catfish begins with the breeding process. Through a system called selective breeding, superior fish are used to provide eggs for hatching. Factors such as growth rate and meatiness of the fish are used to determine which fish are superior. Although catfish have been known to reach a weight of over 100 pounds, the best size for a brood fish seems to be around 3 to 5 pounds.

The fish are placed in ponds less than 6 feet deep called brood ponds. These ponds are used to keep fish that breed and lay eggs. Boxes are supplied for the fish to use as a nest for spawning and are submerged. The female lays several thousand eggs (called roe) at a time (Figure 17-6). After the female lays a mass of roe in the nesting box, the male fertilizes the eggs by depositing sperm into the mass of eggs. In the wild, the male cares for the nest until the eggs hatch.

In aquaculture operations, the producer takes the eggs from the nest soon after they are laid and fertilized. The eggs are placed in a hatching tank where paddles rotate and gently move the water. After the eggs hatch, the small fish, called fry, are placed in long tanks. In these tanks the fry are fed and

FIGURE 17-5
Catfish can tolerate relatively low levels of oxygen.

FIGURE 17-6
The female catfish lays a mass of eggs called roe.

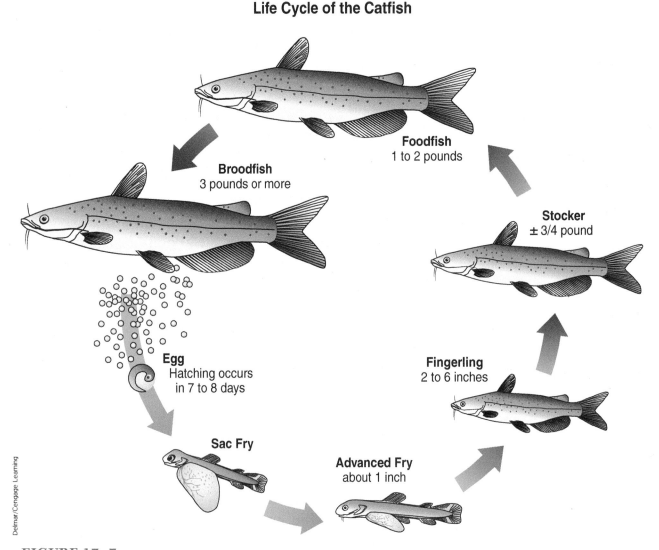

FIGURE 17-7
The life cycle of the catfish.

cared for until they become **fingerlings**. Fingerlings are small fish that are from 2 to 6 inches in length. This size fish is what is used to stock the ponds (Figure 17-7).

The grow-out ponds are about 4 to 6 feet deep and are usually around 1 to 2 acres in area. The smaller ponds make harvesting easier. In these ponds the fish are fed on a protein-rich diet. The feed floats on the water so the fish have to come to the surface to get it. This allows the producer to see the fish each day as they are fed. An alternative is to raise the fish in cages that are submerged in the water. This gives the advantage of having the fish together in one

THE SCIENCE OF AQUACULTURE

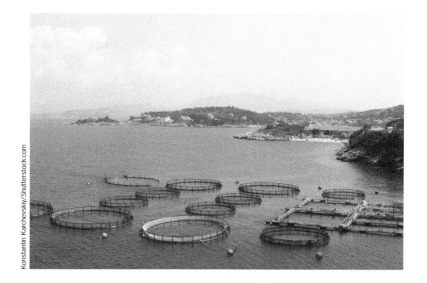

FIGURE 17-8
Raising fish in cages makes them easier to feed and harvest.

FIGURE 17-9
Mechanical aerators sling water into the air to add oxygen.

place. This allows easy feeding, care, and harvesting (Figure 17-8). Also, the cage offers protection from predators.

During the period when the fish are growing, the producer keeps careful check on the pond. Fish, like all other animals, must have oxygen. Their oxygen comes from the oxygen that is dissolved in the water. Water can run out of dissolved oxygen, and if this happens, the fish will die. The producer checks the oxygen level of the water to ensure that the fish have enough. Loss of oxygen occurs most often during hot, dark nights when plants in the water are not giving off oxygen. The producer can put oxygen back into the water by using mechanical aerators. These devices sling water into the air, where oxygen becomes mixed with the water (Figure 17-9).

The fish are harvested when they weigh 1 to 2 pounds. Consumers prefer this size. The fish are harvested using a large net called a seine. The net is dragged through the water to collect the fish. They are then placed in a tanker truck filled with water containing a small amount of salt. The salt has a calming effect on the fish. It is important that the fish reach the processing plant alive (Figure 17–10). Once fish die, they spoil very rapidly. Packers clean and process the fish as quickly as possible. This ensures that consumers receive a fresh, high-quality product.

SUPERVISED AGRICULTURAL EXPERIENCE

Proficiency Awards

The Specialty Animal Production Proficiency Award is one of the many awards available to FFA members. This award is presented to individual FFA members who have effectively produced and marketed such animals as rabbits, fish, dogs, ducks, and bees. The student must have a Supervised Agricultural Experience (SAE) in the selected specialty animal area. The breeding, feeding, and marketing of fish is an excellent project.

Pam's agriculture teacher invited her to join the FFA when she was in the ninth grade. Pam earned the Greenhand degree and attended FFA camp that first year. In her sophomore year, Pam wanted to be a part of the Supervised Agricultural Experience (SAE). After consulting with her parents and agriculture teacher, she decided to grow tropical fish in tanks she kept in her basement. Pam signed up for an aquaculture class and learned how to select, feed, and adjust the environment to produce healthy stock. She purchased the necessary equipment to perform water quality tests and determine weight gains of her fish. In the second year, Pam produced small fish that were sold to local pet stores. Pam's good management practices produced attractive, healthy fish. With the assistance of her agriculture teacher, Pam continued her aquaculture project for three years. Keeping an inventory of equipment, feed, supplies, and the fish was a constant job. Pam used a computer to track the amount of money she spent and made, to calculate each year's net income. At school, Pam won first place in the science fair with a water

THE SCIENCE OF AQUACULTURE

FIGURE 17-10
Fish are delivered to the market alive.

quality project that she tested in controlled tanks with some of her fish. Pam sang in the school chorus and played soccer for the YMCA. By her senior year, Pam had earned her Chapter and State degrees, served as the chapter reporter, and participated in several FFA events. Becoming involved in chapter activities, developing an outstanding supervised experience program, keeping accurate records and inventories, showing the practice of required skills, and preparing a neat and complete application are required for winning proficiency awards. Pam won her chapter proficiency award in Specialty Animal Production and received a medal at the annual FFA banquet. Students who win chapter awards can advance to the district and state competitions. The National FFA Foundation awards plaques and prize money for state and national proficiency winners.

Students with aquaculture projects may qualify for the Specialty Animal Production Proficiency Award. *Courtesy of Blane Marable, Morgan Co. High School, Madison, GA.*

Cool-Water Fish

In the cooler northern portion of our country, trout, Atlantic salmon, and striped bass are raised. These are known as cool-water fish because they thrive in water that is below 65 degrees F. These fish require free-flowing clean water. They must also have a higher oxygen content in the water than catfish. For these reasons, trout and salmon are usually raised in concrete structures called raceways. Here water flows constantly (Figure 17–11). This not only helps keep the water clean but helps keep the oxygen levels high.

The spawning of cool-water fish is different than that of catfish. The procedure usually is aided by a technician. When the female is ready to lay eggs, the technician gently applies pressure to the lower abdomen of the fish. This causes the eggs to ooze out into a container. A similar process is then applied to the male fish, and sperm is collected. The sperm is then applied to the eggs for fertilization. The eggs are hatched in jars in which the water is controlled for temperature and oxygen level. When the fry hatch, they are raised in tanks until they are large enough to stock in the raceways.

FIGURE 17–11
Cool-water fish are raised in raceways of moving water.

Growing Crustaceans

Crustaceans are aquatic animals that have an exterior skeleton (exoskeleton) and three body parts—the head, thorax, and abdomen. Included are shrimp, crayfish, and prawns. These animals have been used as food for thousands of years. Although tons of these animals are harvested in the wild each year, many are raised by producers.

Crayfish are raised in several states. The leading states are Louisiana and Texas. Most are raised in rice fields. The rice fields have to be flooded for a length of time, and during this time the crayfish are raised in the water (Figure 17–12). They eat the decaying

FIGURE 17-12

Crayfish are raised in flooded rice fields.

FIGURE 17-13

Shrimp are also raised in aquaculture operations.

plant material left from growing the rice. They also eat worms and insect larvae. Crayfish are harvested using baited traps that are placed in the water. They are then placed in wet bags that are kept moist and cool and sent to the processing plant. Many of the crayfish are sold live.

Worldwide there are over 35,000 shrimp farms. The major shrimp-producing countries are located in the Eastern Hemisphere (Figure 17-13). China, Thailand, the Philippines, Taiwan, and Japan are major producers. Only about 20 percent of the world's supply is produced in the Western Hemisphere. Central and South America are the leading regions.

Shrimp farms are generally located near the ocean because shrimp have to be raised in salt water. The producers usually make use of the rising tide or floodgates to fill impoundments with salt water. Many of the same technologies used with fish, such as mechanical aeration, are used in shrimp production. The shrimp are harvested by using seines. They are then put on ice and shipped throughout the world.

Prawns are sometimes referred to as freshwater shrimp. These animals have to be grown in very warm water (75 to 95 degrees F). The young (called larvae) have to be grown in brackish water. Brackish means that the water has a salt content lower than ocean water and higher than fresh water. At the proper stage, they are placed in fresh water. Here they are raised until harvest.

Ornamental Fish

One of the fastest growing segments of aquaculture is the growing of ornamental fish, also called fancy fish (Figure 17–14). A favorite hobby of many people is raising tropical fish in home aquariums. These fish come in a wide variety of sizes, shapes, and colors. These fish can be very valuable. Because of their value, extra care and expense can be invested in their

FIGURE 17–14

The raising of ornamental fish is a growing industry.

production. Most tropical fish are raised in what is called a closed-loop system. This system consists of a large tank, a filtration system, and an aeration system. The fish must be kept at a constant temperature and the water must be clean. The spawning process is carefully monitored because many tropical fish eat their young. Once the young are hatched, they are grown to a certain size and shipped to pet stores all around the world.

An ornamental fish called koi is gaining in popularity. This fish is a species of carp and has varying patterns of gold, white, and black. They are used in landscape ponds to decorate yards. These fish can be quite expensive. Breeding fish of the right color and color pattern have been known to sell for millions of dollars.

SUMMARY

Aquaculture is a departure from the traditional concept of agriculture. This relatively new industry will likely grow even more as we enter the twenty-first century. Someday, aquaculture crops such as lobsters, bullfrogs, and other animals now harvested from the wild may be widely produced. Agricultural research will find a way to make this happen.

17 CHAPTER REVIEW

☀ Student Learning Activities

1. Visit the local pet shop and interview the manager. Determine where the shop gets the fish it sells. Ask how they are ordered, shipped, and received. Also, find out how the fish are cared for. Report your findings to the class.
2. Visit the grocery store and compare the price of farm-raised fish with fish caught from the wild. Also compare the cost to beef, pork, and poultry.
3. Go to the library and research a particular type of fish. Determine if the fish might have potential for commercial production. Write your findings in a report.

4. Locate information on how you can grow fish in tanks or aquariums in your classroom. Make a list of all the equipment you will need. What problems are encountered with growing fish in your classroom?

True/False

1. _____ One of the newest and most rapidly developing parts of agriculture is that of growing animals in water.
2. _____ Fish are not very efficient farm animals.
3. _____ Aquatic animals use less energy than land animals.
4. _____ The most important food fish grown in this country is the catfish.
5. _____ Catfish has replaced cotton as the number one cash crop in the Mississippi Delta.
6. _____ Cool-water fish require a low oxygen content in water.
7. _____ Shrimp must be raised in salt water.
8. _____ One of the fastest growing segments of aquaculture is the growing of ornamental fish.
9. _____ Fish is high in protein and low in calories.
10. _____ The fat content of fish is lower than that of red meat.

Multiple Choice

1. The state that leads the nation in catfish production is
   a. Missouri
   b. Mississippi
   c. Florida

2. The production of catfish begins with this process:
   a. breeding
   b. stocking
   c. spawning

3. Small fish that are from 2 to 6 inches in length are called
   a. crustaceans
   b. fingerlings
   c. prawns

4. Most tropical fish are raised in what is called a
   a. closed-loop system
   b. brood pond
   c. raceway

5. Most crayfish are raised in _____ fields.
   a. corn
   b. wheat
   c. rice

6. Catfish eggs are called
   a. roe
   b. seine
   c. prawns

7. Cool-water fish thrive in water that is below
   a. 45 degrees F
   b. 65 degrees C
   c. 65 degrees F

8. Trout and salmon are usually raised in concrete structures called
   a. prawns
   b. raceways
   c. closed-loop systems

9. Crustaceans are aquatic animals that have
   a. endoskeletons
   b. epidermis
   c. exoskeletons

10. The best size for a brood fish is around
    a. 3 to 5 pounds
    b. 7 to 8 pounds
    c. 4 to 6 pounds

Discussion

1. Why is aquaculture such a growing industry?
2. Why did fish farms come into existence?
3. Why are fish considered efficient farm animals?
4. Why are catfish so popular?
5. What are raceways?
6. What are crustaceans?
7. What are the major shrimp-producing countries?
8. What is brackish water?
9. When does the loss of oxygen occur most often?
10. How can oxygen be put back into water?

CHAPTER 18
Companion Animals

Student Objectives

When you have finished studying this chapter, you should be able to:

- Discuss some of the ways companion animals are used to assist humans.
- Explain the benefits of hippotherapy.
- Define zoonoses and list several types.
- Identify several aspects of responsible companion animal ownership.
- List several careers that deal with companion animals.

Key Terms

companion animals
service animals
hippotherapy
zoonoses
parasite

CHAPTER 18

Have you ever considered companion animals to be a part of agriculture? Companion animals are those animals we keep as pets. Often they live with us in our homes and keep us company wherever we go. We would all agree that people who own pets are involved in raising the animals. Companion animals play an important role in the agricultural industry. People have always had a relationship with animals. In the last 50 years, our relationship with many domestic animals has changed. As our society has become more urban, the importance of these animals is measured less by the work they perform for us. Animals still have an important place in the lives of many people, however. They are more than pets; they are friends and companions. It is this special relationship between the animal and its owner that identifies a companion animal.

Companion animals are an important part of our society. In fact, pet ownership is at an all-time high. Almost 60 percent of American households have at least one companion animal. Virtually any kind of pet can be considered a companion animal. Dogs and cats are the most common, but companion animals may also be horses, rabbits, hamsters, gerbils, guinea pigs, and ferrets (Figure 18–1). The companionship bond may also be developed with some unusual pets, such as potbellied pigs, reptiles, or fish. This wide range of pets provides companionship for people of all ages.

FIGURE 18–1
Dogs and cats are the most popular companion animals, but animals such as ferrets are also popular.

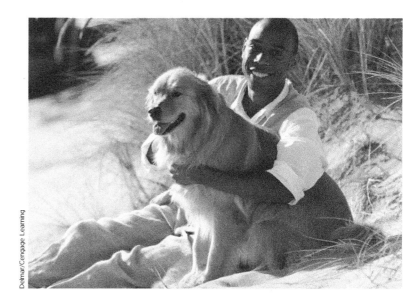

FIGURE 18-2
Dogs are able to express themselves to their owners.

In recent years, cats have outnumbered dogs as the favorite pet. There are about 60 million cats and 52 million dogs in American households. Many cat owners have more than one cat, so although cats are more numerous, dogs are found in more households.

The popularity of dogs and cats as companions is due to a combination of factors. One of these is the ability of dogs and cats to form lasting social bonds with humans. They are able to establish a close relationship with their owners. They can also be house-trained fairly easily. House-training is impossible with many companion animals. Dogs and cats also have the advantage of being large enough for humans to interact with and play with, and yet are small enough to be kept in the house.

The most important reason for the popularity of dogs and cats is their ability to express themselves to their owners. Both dogs and cats have a number of gestures and sounds they use to communicate with their owners (Figure 18-2). People tend to attribute human feelings and emotions to their pets because of this ability to be expressive.

The fastest growing category of pets in the United States is reptiles (Figure 18-3). There are more than seven million pet reptiles living in homes in the United States. Iguanas are the most popular pet among reptiles. These lizards are cat-sized when full

FIGURE 18-3
Reptiles are rapidly becoming some of the more popular pets.

grown. They are clean, odorless, and can be house-trained. This category also includes snakes and tortoises, as well as amphibians such as frogs and salamanders.

Health Benefits

Pets are more than just companions—they are good for people's health. Scientists are now discovering that living with a pet contributes to the physical and mental well-being of humans. Relationships with animal companions appear to be beneficial to humans because they are uncomplicated. Animals are accepting and attentive, and respond to affection. They are not judgmental, they never talk back, they never

CAREER DEVELOPMENT EVENTS

The FFA offers many opportunities for students to develop and display projects. Students even have the opportunity to develop their own programs and the area of companion animal production is well suited for this purpose. If you are involved with the training, grooming, care, or production of companion animals, you might want to organize an event that will showcase your animals.

The first step would be to locate others in your chapter with interest in companion animals. With the help of your advisor, form a committee to discuss the event. Begin by developing objectives and determining what you want to accomplish with the event. The objective might be to have a demonstration of grooming techniques, a competition in grooming, or a competition in the training of companion animals. Use your imagination—there are lots of types of events you could organize.

Once you determine what you want to accomplish, organize the procedures and establish guidelines to govern the event. Get everyone in your chapter involved in the planning and activities, and with your advisor's help, organize the members into committees to get everything done. You might even get the local FFA Alumni Affiliate involved, which offers support for the FFA programs.

criticize or give orders. Pets give people something to be responsible for and make them feel special and needed (Figure 18–4).

Companion animals are a good influence on children of all ages. They help children develop a sense of security. They have been used to encourage shy or withdrawn children to open up. Children who are normally hyperactive often become calmer around a companion animal.

New evidence of the beneficial effects that companion animals have on human health is continually being discovered. We now understand that pets make people feel good, and a sick person who feels good mentally is likely to get better faster. People who have pets report fewer minor health problems, like colds and flu. Studies have also shown that

Students who raise companion animals as an SAE may compete for the Specialty Animal Production Proficiency Award. *Courtesy of Tracy Fera, PET SPAS of America*

As your event becomes successful, you may wish to invite other chapters to participate. This was the way most of the major FFA events began. The FFA is growing and looking for new events, and you just might have an idea that could grow into a national event.

FIGURE 18-4
Pets can make people feel needed.

FIGURE 18-5
Watching fish in an aquarium can help lower a person's blood pressure.

petting a dog or a cat or watching fish in an aquarium can help lower a person's blood pressure and heart rate (Figure 18-5).

Due to the increasing mobility of today's society, many elderly people no longer have family members living close by. Therefore, companion animals play an important role in the lives of a number of older individuals. Many of them would feel isolated and alone without their pets. Pets provide them with companionship and a sense of being valued and needed. Older people may actually live longer, healthier lives because of their relationship with their companion animals.

Today about 50 percent of nursing homes use animals in some capacity. Bird aviaries and aquariums are popular among residents, as well as cats, rabbits, and guinea pigs. It is not recommended that dogs live in nursing homes full-time because they tend to be overfed by the residents, who cannot seem to resist feeding them cookies and treats. Today many volunteers take their dogs to visit the residents of nursing homes and hospitals in their community.

Service Animals

Some companion animals serve the dual purposes of companionship and practical assistance. They are known as service animals. A number of agencies in the United States train assistance dogs. These dogs are trained to give more independence and mobility to people with disabilities. Dogs serve as the eyes, ears, or legs for thousands of people. Training usually takes between four and eight months, depending on the difficulty of the tasks that must be learned and the aptitude of the dog. Although training an assistance dog can cost thousands of dollars, many agencies provide them to people who need them at little or no cost.

The best known example of an assistance dog is the guide dog for the blind. The most commonly used breeds are German shepherds, Labradors, and retrievers (Figure 18–6). To work as a guide dog, the dog has to be exceptional. It has to be able to walk through crowds, climb stairs, ride on elevators, ride on buses and in cars. Most importantly, the dog has to be able to think for itself. It must learn to disobey a command if it could bring harm to its master.

Many of the organizations that train guide dogs have volunteers raise the puppies for the first year. Some of the volunteer puppy raisers belong to the 4-H organization. The volunteers are not expected to train the puppies as guides, but they are required to follow some basic rules. The puppies must be exposed to many of the activities that they will need to handle with ease as a guide.

FIGURE 18-6
The German shepherd is a popular breed used as guide dogs.

FIGURE 18-7
Only about half of the puppies raised to be guide dogs successfully complete the training.

The puppy raisers are encouraged to take the puppies to the mall, to the park, to nursing homes, schools, or any other setting that involves the public. The puppies must be kept on a lead in public, and they must sleep next to the bed of the volunteer, just as they will when acting as guides.

The volunteers are warned not to play ball, tug-of-war, or other games with the puppies that could turn into bad habits when they become guide dogs. The puppy raisers must never feed their pets human food. Guide dogs must not be tempted by the smell or presence of human food, as they may accompany their owners to restaurants. They cannot be jumping up on tables begging for food.

Only about half of the puppies raised to be guide dogs successfully complete the training (Figure 18–7). During training, the dogs learn to work in a harness, and they learn commands like "Forward" and "Find the door." The dogs are trained through repetition and praise. They learn to ignore crowds, noises, squirrels, and cats. The dog must learn that it is as wide and as high as a person. It must be aware of obstructions that it can walk around or under, like awnings or branches that would impede its owner. A guide dog wearing a harness is on duty and should never by petted by other people. When the dog is out of the harness, the dog is like any other family pet.

Hearing-ear dogs are trained to listen for people who cannot. Also called signal dogs, these animals can respond to more than 30 common household sounds like doorbells, telephones, alarm clocks, and fire alarms. They can even be trained to respond to a crying baby. The dogs alert their deaf or hard-of-hearing owners by walking back and forth from the source of the noise to the owner. Signal dogs are taught not to bark when alerting their master to a sound, as they would be unable to hear the barking. The dogs can also be trained to respond to sign language commands. Because the size of the dog is not important, hearing-ear dogs are usually mixed breeds and are often rescued from local animal shelters.

Service dogs are trained to help people who use wheelchairs or have spinal injuries. They are able to respond to more than 40 different commands. Service dogs can open doors, work light switches, pull emergency cords, and pull wheelchairs. Each dog may be trained a little differently in order to address the needs of the individual who will own the dog. Service dogs need to be large and are often retriever breeds.

Another companion animal that has been beneficial to humans is the horse. Horses are used in programs of physical therapy for people with disabilities. This type of therapy, called **hippotherapy**, can be used with people of all ages, but it is especially helpful for physically challenged children. The gait of a horse simulates the motion of humans as we walk. When we walk, our bodies move from side to side and up and down. Riding a horse re-creates that sensation in people who are unable to walk unassisted (Figure 18–8).

Through hippotherapy, physically challenged individuals improve their balance, posture, strength, coordination, and muscle flexibility. In addition to the physical benefits, the riders also gain confidence. Hippotherapy can provide them with a whole new perspective and a sense of freedom. A horse can take its rider where no wheelchair can go. Volunteers are used in the care and maintenance of the horses as

FIGURE 18-8
Horses help people who are unable to walk unassisted. The gait of the horse resembles the motion humans make when they walk.

well as for lessons. Hippotherapy programs offer many opportunities to local agriculture programs and youth organizations.

Diseases and Afflictions

Unfortunately, pets can give us more than companionship. Each year pets pass along infectious diseases to thousands of Americans. There are approximately 30 varieties of pet-borne illnesses, which are called **zoonoses**. Zoonoses are diseases and infections that can be transmitted from animals to humans. They can be passed on through direct contact with the animals or acquired indirectly through contact with animal feces or other contaminants. However, most diseases passed from animals to humans are easy to avoid and are treatable. Good hygiene and safe handling procedures should always be practiced when working with animals (Figure 18-9). Cats and dogs are responsible for the majority of zoonoses, but birds, fish, and turtles are also culprits.

Rabies is the best known and the most feared example of a zoonosis. It is contracted through the saliva of rabid animals. Although this disease is rare in pets, rabies is increasing in the wild animal population. Because pets may come into contact with wild animals, they should be vaccinated against the disease.

The common roundworm of dogs is a **parasite** that can infect humans. A parasite is an organism that lives off of or at the expense of another organism. The parasite is transmitted through contact with animal feces or with contaminated soil. Children playing in areas frequented by dogs are especially at risk. Dog owners should make sure their pets are regularly dewormed (Figure 18-10).

Toxoplasmosis is a parasitic disease that can be transmitted by contact with cat feces. Toxoplasmosis can be especially dangerous to pregnant women, so most veterinarians recommend pregnant women not clean cat litter boxes.

FIGURE 18-9
Good hygiene and safe handling procedures should be practiced when working with animals.

FIGURE 18-10
Dog owners should make sure their pets are regularly dewormed. The medication can be given by injection or by mouth.

Psittacosis, or parrot fever, is a disease transmitted by parrots, budgerigars, and other related caged birds. Humans can be infected by contact with the feces of contaminated birds. In handling birds and cages, dust masks or protective face shields should be used.

Ringworm is not a worm but a fungus that results in a skin aggravation in humans. It is primarily passed to humans by kittens and puppies. The animal appears unaffected because the fungus infects only the animal's fur. It is passed to you when you handle your pet. It is more common in children because adults seem to become more resistant to it with age.

Rocky Mountain spotted fever and Lyme disease are tick-transmitted diseases that can affect both humans and animals (Figure 18-11). Ticks are found in grassy, wooded areas, or they can be brought into the household by dogs and cats that have been outside. Both diseases are treatable with antibiotics. Normal grooming of animals after they have been outside will help locate and eliminate ticks.

Infections due to animal bites and scratches are another concern. The potential for infection varies. Less than 5 percent of dog bites become infected, but up to 50 percent of cat bites do. Cat scratch fever is associated with cat scratches or bites. This disease

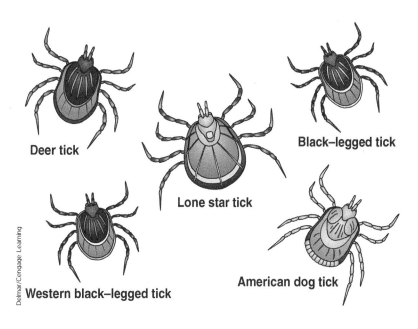

FIGURE 18-11
Ticks can transmit Lyme disease.

is not serious and can be treated with antibiotics. Safe handling techniques are an important measure to prevent bites and scratches, as well as injury to the animal. Prompt and thorough washing of pet bites and scratches with soap and water is always important.

Allergies are probably the most common afflictions that occur when humans have contact with animals. Many people develop allergies to animal hair and their dander, or flaking skin. Allergies cause hay fever–like symptoms that can occur in both children and adults.

Industry

One measure of the value people place on pets is the amount of money they are willing to spend on them. Americans spend over $20 billion a year on companion animals! Dog and cat food purchases account for about $8 billion of the total spent on pets. By comparison, people spend approximately $1 billion on baby food. The companion animal industry contributes to the economy through the sale of pets, pet food, accessories, and veterinary care (Figure 18–12).

The pet food industry utilizes many of the by-products and surpluses of the human food industry. The main ingredient in dry dog foods is grain, soybean meal, or wheat millings. The main ingredients in canned pet foods are meat by-products, which may include the waste products of meat that was processed for human consumption. The composition of pet food is carefully formulated to meet the nutritional needs of the animals. Canned, semimoist, and dry foods are equally nutritious, but the canned varieties generally contain a higher percentage of protein and fat. Consequently, pets should be fed smaller amounts of canned food or mix dry foods with canned.

Cats have special dietary needs. Pet foods formulated for cats have a much higher protein and fat content than any type of dog food (Figure 18–13).

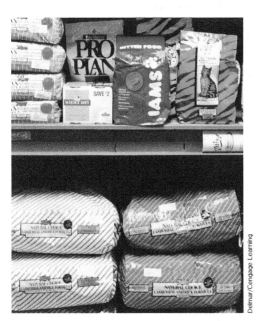

FIGURE 18–12

The pet food industry produces a wide variety of pet foods.

COMPANION ANIMALS

FIGURE 18–13
Cats require a diet higher in protein and fat than dogs.

Dogs often seem to prefer cat food because of the high fat content. For healthy, active dogs, nibbling a little cat food is not a problem, but they will become obese if allowed to eat it regularly. In contrast, cats should not be allowed to eat dog food because it does not contain all of the nutrients that a cat needs to stay healthy.

The supply of pet accessories is also an important contributor to the economy. A companion animal may need any number of essential items such as a feeding bowl, dish, or trough; a cage, hutch, or kennel; and chew bones, scratch poles, and toys. Many pet owners also indulge in luxury items such as heated pet beds, jewelry, pet clothing, and car seats.

Health Care

Americans spend over $10 billion every year on the health care of their companion animals. There are approximately 40,000 veterinarians in the United States; one-third of them treat small animals exclusively (Figure 18–14). Like all veterinarians, those who specialize in the care of companion animals perform a wide variety of tasks every day. They treat animal injuries, set broken bones, immunize healthy animals against disease, and perform surgery.

FIGURE 18–14
One-third of all veterinarians in the United States treat small animals exclusively.

Veterinarians today can successfully perform hip replacements and kidney transplants on companion animals. They have also performed balloon angioplasty to open clogged arteries, open-heart surgery, and dental surgery. In recent years, medical care for animals has become as highly technical as medical care for humans. In fact, new medical procedures are perfected on animals, so in some instances animals may get even more advanced care than humans. However, the costs of these health care advances may be more than some pet owners are willing or able to bear.

Veterinarians stress preventive measures when they counsel pet owners. They encourage the pet owner to include vaccination programs and regular dental exams in their pet's health care plan. According to veterinarians, one of the most common problems they see in dogs and cats today is obesity. Half of American dogs and almost one-fourth of cats are obese. The animals suffer from overfeeding and a lack of exercise. Some animals simply do not get enough attention. When animals get bored, they have a tendency to eat too much, just like people. A sound diet and daily exercise routine should also be a part of the pet's overall health care plan.

Responsible Ownership

Responsible ownership of a companion animal requires a long-term commitment and a realistic assessment of the needs of the pet. A number of factors should be considered by the prospective pet owner that will contribute to the well-being of the pet. The first is the amount and kind of space in which the pet will live. A lively hunting dog does not belong in an apartment (Figure 18–15). Likewise, an easily frightened pet should not be kept in a household with several lively, noisy children. No dog should be tied to a tree and left alone all day, every day. Animals that are forced to live in environments that do not suit their natures may exhibit behavioral problems.

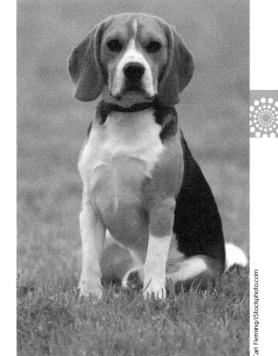

FIGURE 18-15
A lively hunting dog does not belong in an apartment.

FIGURE 18-16
Small animals such as mice do not require much space.

Although many households want to have dogs, busy families that do not have the time or space to devote to a dog have many other options to consider. Rabbits, hamsters, mice, and guinea pigs can be affectionate, cuddly pets and do not require the time or the space of a dog (Figure 18-16). Reptiles, such as geckos, monitors, and iguanas, are not soft and cuddly, but many of them do seem to enjoy contact with people, probably because of human body heat. Reptiles do not require much space and may even be kept in apartments that do not allow cats and dogs.

Responsible owners of companion animals also need to consider the problem of pet overpopulation. Millions of unwanted and homeless animals are born each year. This overpopulation creates problems for society and leads to suffering for many companion animals. To solve this problem, the number of puppies and kittens born must be reduced.

At least 10 million dogs and cats are destroyed, or euthanized, each year by animal pounds and shelters. Future pet owners are encouraged to consider rescuing an animal by adopting a pet from their local animal shelter. It is the responsibility of current pet owners to help control this problem by neutering and spaying their pets.

Spaying and neutering are surgical procedures that prevent animals from reproducing. Spaying is performed on females and neutering is performed on males. The animals are anesthetized so they do not feel any pain during the surgery. They may experience some discomfort for a day or two after the procedure.

Spaying and neutering a pet can result in a healthier, happier animal. Spayed females are less susceptible to some infections and cancerous growths. Neutered males are less likely to display aggressive behaviors or to roam.

SUMMARY

Companion animals represent a large and growing segment of animal production. The many segments of this vast industry have a positive effect on the economy. It is different from most of the agricultural industry in that it is not involved with the production of food and fiber. It does, however, provide a valuable product for people who enjoy animals as companions.

CHAPTER REVIEW

Student Learning Activities

1. Visit the local pet store and interview the owner as to where the animals come from. Research how the animals are produced.
2. Make a list of all the money spent on your pet during the last year. Determine how many different industries were supported by the expenditures.
3. Visit the local humane society and determine how your class can help in educating people about animal care.
4. Go to the Internet and research a type of companion animal. Determine how long it has been in domestication and the country of origin. What are some of the popular breeds?

True/False

1. _____ There are more cats than dogs in the United States.
2. _____ Dog food can also be fed to cats with no ill effects.
3. _____ A companion animal may be a snake.
4. _____ A zoonosis is a disease that is passed from a caged animal to a companion animal.
5. _____ Lyme disease is contracted through contact with cat feces.
6. _____ Rabies is relatively uncommon in household pets.
7. _____ Veterinary technology tends to lag far behind human medical technology.
8. _____ Most dog bites become infected.
9. _____ The fastest growing category of pets in the United States is reptiles.
10. _____ Spaying is a surgical procedure performed on female animals.

Multiple Choice

1. Ringworm is caused by
    a. parasites
    b. fungus
    c. worms

2. Signal dogs are trained to assist
    a. people in wheelchairs
    b. the blind
    c. the deaf

3. The main ingredient in dry dog food is
    a. grain
    b. chicken
    c. meat by-products

4. Hippotherapy is particularly beneficial for
    a. the elderly
    b. physically challenged children
    c. the deaf and hard-of-hearing

5. Volunteers who raise puppies for guide dog programs should try to expose the puppy to
    a. games like tug-of-war
    b. human food
    c. crowds of people

6. Neutering is a surgical procedure that is performed on
   a. male animals
   b. reptiles
   c. female animals

7. Service dogs are trained to help
   a. the blind
   b. the sick
   c. people in wheelchairs

8. The best known and most feared zoonosis is
   a. rabies
   b. Lyme disease
   c. ringworm

9. It is not recommended that dogs live in nursing homes full-time because
   a. they tend to be aggressive toward elderly people
   b. they tend to be overfed by the residents
   c. companion animals are not beneficial for elderly people

10. Psittacosis is transmitted by
    a. contact with the feces of some caged birds
    b. ticks
    c. parasites

Discussion

1. Why is spaying or neutering a pet an important part of responsible pet ownership?
2. How does the companion animal industry contribute to the economy?
3. What are the differences between dry and canned pet foods?
4. Why are dogs and cats the most popular companion animals?
5. What is a companion animal?
6. How do companion animals affect the health of humans?
7. Why are companion animals beneficial to elderly people?
8. What are zoonoses?
9. How do assistance dogs help people?
10. What is hippotherapy?

CHAPTER 19
Preserving Our Food Supply

Student Objectives

When you have finished studying this chapter, you should be able to:
- Discuss the need for food preservation.
- List the major ways food is preserved.
- Discuss modern trends in food consumption.
- Explain how food can become unsafe to eat.
- Discuss how the USDA helps to ensure a safe food supply.

Key Terms

drying
jerky
pickling
salting
canning
blanching
freeze-drying
mad cow disease
USDA
pesticide residues
E. coli bacteria

CHAPTER 19

One of the most important segments of our huge agricultural industry is concerned with preserving our food. No matter how much food we produce, it is of little use if it does not reach consumers in an edible form. As you know from studying the science of living organisms, plants and animals begin to decompose when they are no longer living. Microorganisms such as bacteria feed on plant and animal tissue as a means of returning dead organisms back to the soil. This happens with our food crops also. When plants and animals are harvested for food, the tissues within the food begin to break down. Without a means of preserving the food, it would spoil before it could be eaten.

While some of our food is sold fresh, the vast majority of food products are preserved in some manner. Many of our green leafy vegetables are sold fresh, but these foods do not last long in the fresh state (Figure 19–1). Fruits such as apples, bananas, and grapes are bought fresh but still need refrigeration to make them last. Also, meat such as beef, pork, and chicken is sold fresh, but it must be eaten within a few days or it will spoil. The percentages of these foods that are sold fresh keep declining. In our modern era, most households do not have someone whose sole job is that of keeping house, shopping, and cooking. People tend to want to visit the grocery store less often than in years past. This means that food must last

FIGURE 19–1

Many of our green leafy vegetables are sold fresh, but these foods do not last long in the fresh state.

Courtesy of USDA

FIGURE 19-2

With the advent of the microwave oven, there is great demand for food that can be taken from the freezer and cooked in a few minutes.

longer than just a few days. Another trend is toward foods that are not only frozen but are also cooked. With the advent of the microwave oven, there is great demand for food that can be taken from the freezer and cooked in a very few minutes (Figure 19-2).

Methods of Preserving Foods

Before methods of preserving foods were developed, people feasted when food supplies were plentiful. For example, when a large animal was killed, there might be plenty of meat for everyone. However, this only lasted a few days, because the meat began to spoil. When fruits were ripe or vegetables were at their peak, people ate well. During the colder months, when the fruits and vegetables disappeared, there were none available. Another problem encountered was that of travel. When people traveled, they most often had to take food with them. If the trip was long, fresh food would spoil before the journey ended. This meant that food needed to be preserved to last the entire journey. Perhaps the greatest problem was that of feeding an army, especially if that army traveled. Not only did food have to be preserved, it also had to be preserved so that it would taste good when it was eaten. These problems led to the development of food preservation. Basically, all food preservation methods have at least one thing in common—they slow down or stop the growth and reproduction of microorganisms such as bacteria (Figure 19-3).

FIGURE 19-3
Bacteria such as these cause food to spoil.

Drying

The first method of food preservation was probably drying. People noticed that as plant seeds matured, they became dry. This allowed the seeds to survive during the winter months. People began to use the sun and fires to slowly dry foods. Fruits were dried and preserved. Meats were cut into thin slices, which were hung over a fire until they dried. Many Native Americans in the Pacific Northwest survived by catching and drying salmon that were caught during a certain time of the year when the fish came up small streams.

Drying preserves food because the moisture is removed from the plant or animal tissues. In order to survive, microorganisms must have moisture. Without moisture, they cannot grow and reproduce. This method worked so well that it has survived from prerecorded history until the present. We still use this method to preserve food. Dried fruits such as dates, prunes, and raisins are readily available in the grocery store (Figure 19–4). Dried meats such as jerky can be found in most convenience stores.

Pickling and Salting

Even though this method of food preservation goes back hundreds of years, it is still used today. Almost all of us enjoy pickles on our hamburgers and

PRESERVING OUR FOOD SUPPLY

FIGURE 19-4

Drying preserves such foods as raisins. Note the difference between the raisins after drying and the grapes before drying.

FIGURE 19-5

Pickling preserves food by using vinegar as a medium.

sauerkraut on our hotdogs. **Pickling** preserved both of these foods. Vinegar has been used as a medium for pickling for centuries (Figure 19–5).

It works because vinegar is so acid that bacteria will not live in the liquid. Sauerkraut is preserved because of the addition of salt and because the cabbage undergoes a fermentation process very much like farmers use to create silage for livestock. This process also makes the sauerkraut more acid and aids in preservation.

Like pickling, salt has been used to preserve food for almost as long as we have recorded history. High concentrations of salt cause bacteria to die by

FIGURE 19-6
Hams are preserved by salting and smoking. This food is popular all over the world.

drawing the moisture out of organisms. This process is called osmosis. Meats and fish have been preserved by this method, which is called salting. In the early days of our country, settlers used salt to cure their meat. Buildings known as smokehouses were constructed to hang the meat. Salt was forced into the meat and it was smoked to prevent insect damage. Even today, salt cured hams and fish are very popular (Figure 19-6).

Canning

We all use cans of fruit, vegetables, meats, and a wide variety of prepared foods. This process works by placing foods in metal cans or glass jars. The cans or jars are then sealed so that nothing can get in or get out until the container is opened. Subjecting the filled containers to a high level of heat kills the bacteria and other organisms (Figure 19-7). This process, called canning, is done in what is called a retort. This device puts the cans or jars of food under pressure supplied by steam. Most food contains water. If the cans of food were heated high enough to kill all of the microorganisms, the water in the cans would turn to steam and build enough pressure to explode. As you know from science class, water evaporates (turns to vapor) at 212 degrees F at atmospheric pressure. However, if pressure is added, the temperature needed to change water to vapor (steam) increases. Putting the containers of food under pressure allows them to reach high enough temperature to kill bacteria and other organisms in the food.

Refrigeration

The invention of refrigeration brought about a revolution in the way foods are preserved. When the temperature is lowered, microorganisms cannot reproduce as rapidly as they can at warmer temperatures. Fresh fruits and vegetables as well as all types of meat can be preserved for days under refrigeration (Figure 19-8).

FIGURE 19-7
In the canning process, food is sealed in a can and sterilized using heat.

FIGURE 19-8
Meat can be preserved for days using refrigeration.

This means that these food products can be shipped all across the country on refrigerated trucks or railcars. Oranges grown in Florida can be shipped all the way to Canada and beef slaughtered in Iowa can be shipped to both coasts. Because of refrigeration, we can enjoy a much broader array of different foods.

Freezing

A common way of preserving food is by freezing. While lower temperatures (refrigeration) slow the growth of microorganisms, very low temperatures (below 0 degrees F) keep bacteria from growing and reproducing, so the food does not spoil. Fruits, meats, vegetables, fish, and other forms of food are harvested, processed, and quickly frozen. Some foods such as vegetables undergo a process called blanching. This procedure involves immersing the vegetables in hot water then rapidly submerging them in ice water. This stops enzymes in the vegetables from causing changes in the plant tissues. Enzymes are substances that cause or speed up a chemical reaction.

Once the food is frozen, bacteria cannot reproduce in the low temperature (Figure 19-9). When the food is thawed, the bacteria can then begin to reproduce and the food will spoil unless it is prepared and eaten within a short period. This process also allows food to be cooked and then frozen. Microwave ovens

FIGURE 19-9
Once food is frozen, bacteria cannot reproduce.

are often used to turn precooked frozen foods into ready to eat meals. This is becoming an increasingly more popular way to buy and prepare meals.

AGRISCIENCE FAIR

The FFA offers students many opportunities to develop and display projects. The National FFA Agriscience Fair recognizes students who are studying the application of scientific principles and emerging technologies in agriculture. Much like the International Science and Engineering Fair, the Agriscience Fair reflects an agricultural theme. Participation begins at the local chapter level and advances to state and national levels. Competition is open to all FFA members in grades 7–12.

One of the major goals of the Agriscience Fair is to provide students with an opportunity to use the scientific process. They can demonstrate and display agriscience projects while reinforcing skills and principles learned in agriscience courses. Award winners are also recognized for their efforts. Certificates, medals, ribbons, plaques, and even scholarships and cash awards may be given to division winners in each category.

There are five categories for the National FFA Agriscience Fair: Biochemistry/Microbiology/Food Science, Environmental Sciences, Zoology (Animal Science), Botany (Plant/Soil Science), and Engineering (Mechanical/Agriculture Engineering Science). Each member or team of two members may enter only one project. The participants are required to meet with the judges to explain their project and will be judged on how well the scientific method was followed,

PRESERVING OUR FOOD SUPPLY

A variation of freezing foods is the process of *freeze-drying* foods. This method combines the concepts of both freezing and drying. Remember from your science class that water comes in three states—liquid, gas, and solid. At room temperature it is a liquid. At 32 degrees F, it begins to turn to a solid (ice). At 212 degrees F, it begins to turn to a gas (vapor or steam). Also remember that these changes take place at these temperatures at atmospheric pressure. If water is put under pressure, it has to get much hotter to turn to vapor. If pressure is removed in a vacuum, water will turn to vapor at a lower temperature. In the freeze-drying process, frozen food is placed under a tight vacuum. Even though it is

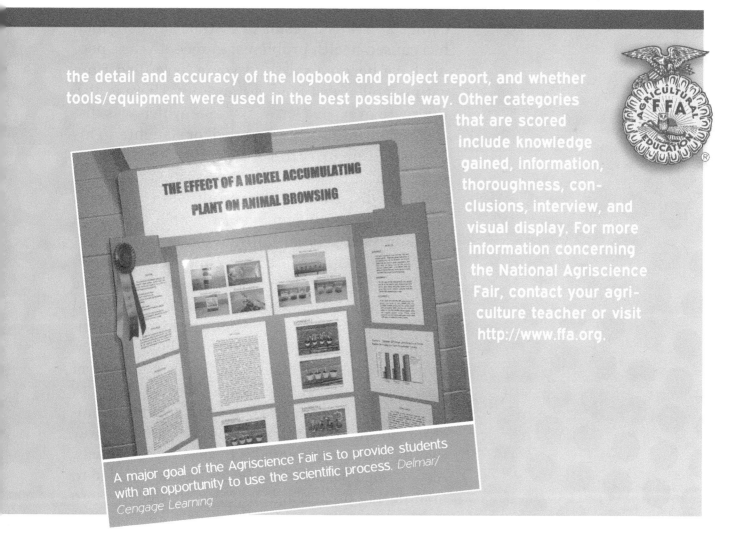

the detail and accuracy of the logbook and project report, and whether tools/equipment were used in the best possible way. Other categories that are scored include knowledge gained, information, thoroughness, conclusions, interview, and visual display. For more information concerning the National Agriscience Fair, contact your agriculture teacher or visit http://www.ffa.org.

A major goal of the Agriscience Fair is to provide students with an opportunity to use the scientific process. Delmar/Cengage Learning

FIGURE 19-10
Freeze dried foods are well preserved and lightweight.

frozen, the water in the food will evaporate, even at temperatures below 0 degrees F. Another way to explain the process is that it is the reverse of the principle used in canning. This process results in a well-preserved food product that weighs relatively little (Figure 19-10). Backpackers use freeze-dried foods extensively.

Food Safety

Americans enjoy the most abundant, nutritional, and safest food supply of any people in history. Almost all consumers are concerned about the food they eat. After all, our health and well-being depends on the wholesomeness of the food we eat. However, some people are overly concerned about the safety of our food supply. This concern comes about as a result of instances where food has caused health problems. In recent years, people have become sick and some have died from eating food at fast food restaurants. There is also the scare over mad cow disease, which was discovered in an animal scheduled for slaughter. This disease occurs in cattle. A type of protein called a prion that attacks the nervous system causes it. Mad cow disease can be transmitted to humans if they eat poorly cooked beef from an animal with the disease. It is almost always fatal.

While all these are real concerns, the overwhelming majority of our food is safe and wholesome (Figure 19-11). Generally all problems with food can be traced to one of three areas: food contamination, food spoilage, or improper food preparation. Food may be contaminated in a variety of ways. For example, insects may have gotten into the food product. Insects can carry a variety of microorganisms that transmit disease. Another way that food is contaminated is by pesticide residuals. Also, foreign particles can be accidentally added to food during processing. Anything that comes into

PRESERVING OUR FOOD SUPPLY

FIGURE 19-11
The overwhelming majority of our food products are safe and wholesome.

contact with food and interferes with the purity of the food product causes contamination. Food contamination is generally considered to be a relatively minor problem because it happens so infrequently. Government agencies under the direction of the United States Department of Agriculture (USDA) monitor how all food products are processed, preserved, packaged. Food products are constantly inspected for contamination of any kind.

Perhaps the most closely monitored of all food products is meat. Trained meat inspection experts carefully inspect each and every animal that is slaughtered (Figure 19-12). The inspectors look for signs of animal diseases or anything else that might compromise the purity of the meat product. All foods are monitored for pesticide residues. Residues are the small amounts of a pesticide or other materials left on crops or animals at harvest. Almost all crops and animals are subjected to pesticides at one time or another. The USDA restricts the type of pesticide that can be used on food plants and animals. The amount of pesticide on food products is also checked to ensure that whatever residues left on the products are not harmful.

A common contaminant on meat is E. coli bacteria. Sometimes during the slaughtering process, bacteria from animal feces come into contact

FIGURE 19-12
Meat products are closely monitored and inspected.

FIGURE 19-13

To ensure safety, poultry should be cooked to 180 degrees F.

with the meat. This bacteria can make us ill if it is consumed with the meat. This means that meat must be thoroughly cooked to kill any bacteria on the meat. This is particularly true of ground beef that we eat as hamburgers. Because the meat is ground up into many tiny particles, there is a lot of surface area that may be contaminated with *E. coli* bacteria. Meats such as steaks and roasts should be cooked until the center of the meat reaches at least 145 degrees F. Ground beef should reach a temperature of at least 165 degrees F and poultry should be cooked to 180 degrees F (Figure 19-13).

Care should also be taken in the way meat is prepared for cooking. If the meat is frozen, it should be thawed in the refrigerator or in a microwave oven. It should never be thawed at room temperature. Thawing at room temperature allows the rapid buildup of bacteria because of the warm temperature and the time it takes to thaw. Dishes, countertops, and utensils that come into contact with raw meat should be thoroughly washed in hot soapy water. Also, hands should be washed after handling meat.

Foods can also spoil after they are packaged. In fact, all food products will eventually spoil over time. Frozen foods, canned foods, dried foods, and pickled foods will sooner or later become inedible. Preserved

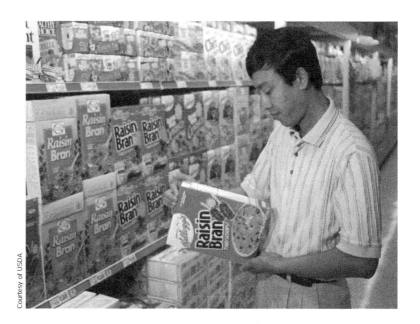

FIGURE 19-14
Preserved food products have an expiration date printed on the package that indicates the last day the product can be sold.

food products have an expiration date printed on the package that indicates the last day the product can be sold (Figure 19–14).

For example, a jug of milk may have a date printed on it. This does not mean that the milk is spoiled after the day indicated. This means that the grocer cannot sell the milk after that day. Sometimes the label designates the day when food loses its freshness. A box of cereal may have such a date. The cereal may be safely eaten after that day, but it might not taste as fresh.

Rarely, a food may spoil before the expected time. This is usually because of improper preservation or packaging. Food that spoils in this manner is dangerous. The same bacteria that cause the spoilage can also make us ill. Beware of any food package that seems swollen or cans of food that have a domed top. This is an indication that bacteria have rapidly reproduced. These bacteria produce wastes and gases that cause the sealed can to expand and bulge out at the top. These cans should be thrown away and not opened. Spoiled food may also have an unpleasant odor or off-flavor. If you notice a bad odor or bad taste in food, do not eat it. Any spoiled food should be disposed of immediately.

CHAPTER 19

SUMMARY

American farmers produce an abundance of food for our population. Modern techniques allow us to preserve the food so it can be eaten year-round. Regulations by the USDA, proper storage methods, and proper cooking practices allow us to enjoy the safest, most wholesome food supply in the world. Following proper procedures when preparing foods and watching for danger signs will help keep our food safe.

CHAPTER REVIEW

Student Learning Activities

1. Make a list of all the food in your home. Beside the name of each food, write down the method that was used to preserve the food. Which method do you think best preserves the food's flavor? Which of the methods are the easiest to prepare for cooking?
2. Interview an elderly person about how food was preserved when he or she was young. Have there been any changes? What do you think brought about the changes?
3. Visit a grocery store and make a list of all the different ways meat is preserved. Why are there so many different ways? Which method do you like the best? Why?
4. Conduct an Internet search and locate information on new techniques of preserving food. Which do you think has promise? What are some of the problems? Do you think these problems can be solved?

True/False

1. _____ Almost all of the food we eat is sold fresh.
2. _____ People visit the grocery store less often than in the past.
3. _____ The first method of preserving food was probably drying.
4. _____ Dried foods are no longer available.

5. _____ The use of pickling as a method of preserving food is only about 20 years old.
6. _____ Salt cured hams are popular today.
7. _____ Refrigeration kills all of the bacteria in food.
8. _____ Americans have the safest, most wholesome food supply in history.
9. _____ Rare hamburgers are tastier than well cooked hamburgers and should be eaten often.
10. _____ A can of food with a domed top is a sign of food spoilage and should not be opened.

Multiple Choice

1. A dried meat product is known as
   a. potted meat
   b. sausage
   c. jerky

2. A common substance used in pickling food is
   a. vinegar
   b. alcohol
   c. olive oil

3. A device used to put cans of food under high pressure and heat is called a
   a. vacuum chamber
   b. retort
   c. cooker

4. Blanching is a process that kills
   a. bad taste
   b. microorganisms
   c. enzymes

5. An advantage of freeze dried foods are that they are
   a. lighter in weight
   b. less expensive
   c. easier to process

6. Mad cow disease is caused by a
   a. bacterium
   b. prion
   c. virus

7. *E. coli* bacteria is a big problem with
   a. vegetables
   b. fruits
   c. meats

8. Poultry should be cooked to a temperature of at least
   a. 125 degrees F
   b. 460 degrees F
   c. 180 degrees F

9. The expiration date on milk means
   a. that the milk spoils on that day
   b. that the milk cannot be sold after that day
   c. nothing

10. Spoiled food can be detected by
    a. a bad taste
    b. a bad smell
    c. both a and b

Discussion

1. What causes food to spoil?
2. Why do people visit the grocery store less often than in the past?
3. What problems did the invention of food preservation solve?
4. List some modern foods that are preserved by drying.
5. Explain why the invention of refrigeration was so important to the food industry.
6. Why is it important to put frozen food under a vacuum in the freeze drying process?
7. List three ways food can become contaminated.
8. Explain why it is so important to thoroughly cook meat.
9. Where does *E. coli* bacteria come from?
10. What causes a can of food to have a domed top?

CHAPTER 20
The Ethical Treatment of Animals

Student Objectives

When you have finished studying this chapter, you should be able to:

- Distinguish between arguments of animal rights and animal welfare proponents.
- Explain why producers treat their animals well.
- Discuss why people may feel that agricultural animals are mistreated.
- Give reasons why confinement operations are not cruel to animals.
- Discuss why the use of management practices is humane.
- Discuss the use of animals for experimentation.

Key Terms

animal welfare
animal rights
confinement operations
farm factories
farrowing crates

People have always used animals. Since humans first appeared on the Earth, they have hunted and killed animals for food, shelter, and clothing. Very early they began to raise animals in addition to hunting them. From the earliest recorded history to today, people have controlled the lives of animals they raised.

In years past, most people were familiar with farm animals and understood how they were raised and sold for slaughter. Today in the United States, only a small percentage of the people really understand the raising of farm animals (Figure 20–1). This creates an atmosphere where much of what happens can be misunderstood. Some people see the raising and management of animals to be cruel. They feel it is unjust to grow and slaughter animals for our use.

There are at least two groups with different lines of thinking concerning the well-being of animals. The first is generally referred to as those who endorse animal welfare. These people feel that it is all right to raise animals for human use. They say that animals should be cared for properly and that the animals should be comfortable in their surroundings. The other group, referred to as animal rights advocates, feel that animals should be free and that they have as much right to live as do humans. They uphold the notion that all animals that are kept in pens, cages, or pastures should be set free.

FIGURE 20–1
Today relatively few people really understand how agricultural animals are raised.

Animal Welfare

Producers of agricultural animals would be the first to agree that animals should be treated well. The problem comes about when animal welfare activists disagree with producers over the question of what is proper treatment. Producers know that animals have to be cared for properly. Animals that are abused or mistreated in any way will not grow or produce very well. Producers raise animals for profit. Animals that are miserable in their surroundings do not eat, grow, and reproduce efficiently. Unless animals grow fast and reproduce, the producers cannot make much profit from the animals. Because of this, it is in the best interest of the producer to treat the animals well (Figure 20–2). In addition, most producers raise animals for a living because they like animals. For thousands of years, humans have almost literally lived with their animals to care for and protect them. Sheepherders have always had a reputation for caring for their flocks, often at the risk of their lives (Figure 20–3).

Millions of dollars and countless hours of time have gone into research to determine what makes animals comfortable. Modern facilities have been designed to make animals as comfortable as possible. Housing is built so that the space for the animals is not too warm or too cold. Producers realize that if animals are too hot or too cold, they are stressed. Stress causes animals to use energy that could otherwise be used for putting on weight.

FIGURE 20-2
Producers know that to make a profit animals have to be well cared for.

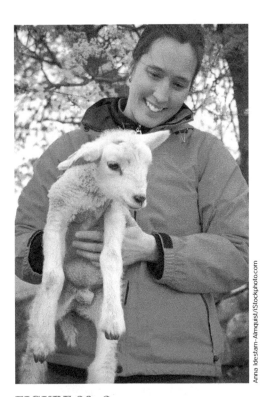

FIGURE 20-3
Sheepherders have always had a reputation for providing complete care for their flocks.

Confining Animals

Animal welfare proponents sometimes object to putting animals into confinement operations. They say it is wrong to give animals only a limited space to live (Figure 20-4). They point out examples of pigs being raised their entire lives in small pens, laying hens being confined to cages, and sows being kept in farrowing crates. The animal welfare activists contend that animals are being mass produced in what they call corporate-owned farm factories.

Producers point out that most of the animals raised are produced on family farms and not on corporate-owned farms. They also argue that facilities are designed specifically for certain animals. For example, a hog house is designed for hogs. This design takes into account the amount of space that a hog needs to be comfortable. It also accounts for the nature of the animal. Pigs are sociable animals and like to be around other pigs. To separate them into isolated spaces where they are by themselves would stress the animals. Research has shown that most animals do better when raised with other animals of their own kind.

FIGURE 20-4
Some people feel it is cruel to raise animals in a confinement operation.

THE ETHICAL TREATMENT OF ANIMALS

FIGURE 20-5
Sows are kept in farrowing crates to protect the baby pigs.

The use of farrowing crates is another example of a facility designed and used for the betterment of the animal. To those who do not understand the purpose and use of farrowing crates, it might look cruel. Sows are placed in the crates about a week before they give birth. They remain in the crate for about a week after the pigs are born. The sow cannot turn around in the crate and has only room to lie down. The purpose is to prevent her from crushing her piglets as she lies down (Figure 20-5).

The crate is designed so that the sow has to lie straight down instead of flopping onto her side. This slower movement allows the little pigs time to move before the mother lies on them. The sows are given plenty of feed and water and the area is kept dry and warm. The sows seem to be in no discomfort and many young pigs are prevented from being crushed to death.

Animals have been selectively bred to live in confinement. This means that they are quite different from their ancestors that roamed freely. Layers that are kept in cages would probably not do well roaming free (Figure 20-6). The cages are designed for cleanliness and comfort for the hens. They have an abundant source of feed and water and the temperature in the laying houses is carefully controlled.

Animals that are kept in confinement can receive better care than animals that roam free. Parasites have always been a problem to animals. They cause

FIGURE 20-6
Hens that are kept in cages have been bred for this type of life.

CAREER DEVELOPMENT EVENTS

You already know that agriculture is more than farming. Any students interested in a career in livestock should strengthen their knowledge and experience by becoming active in the FFA. The FFA develops competent and aggressive leaders who learn how to speak out and debate issues such as animal welfare. Agriculture education and the FFA provide hands-on experience in practicing skills of health management with animals. This experience gives you a better understanding of animal welfare issues.

Animal welfare has become an issue of debate in local, state, and national agricultural meetings around the world. To speak out and present your views requires the use of correct parliamentary procedure. All FFA members should learn how to conduct and participate in local meetings. Parliamentary procedure teaches the student an orderly method of conducting business that ensures that all sides of an issue are treated fairly. Most FFA chapters conduct a local Parliamentary Procedure Event where six students make up a team. The team is given tasks to perform using correct parliamentary procedure. In preparing for the event, team members must learn proper use of the gavel, voting procedures, order of motions, and other stress and discomfort and may even shorten the life of the animal. When animals are kept in confinement, producers make certain that the animals are kept free of both internal and external parasites. In addition, the animals do not have to worry about predators, like animals in the wild or those running free. The producers see the animals every day and care for their needs.

Management Practices

Another problem perceived by animal welfare activists is that of the management practices used by producers. They say that practices such as branding,

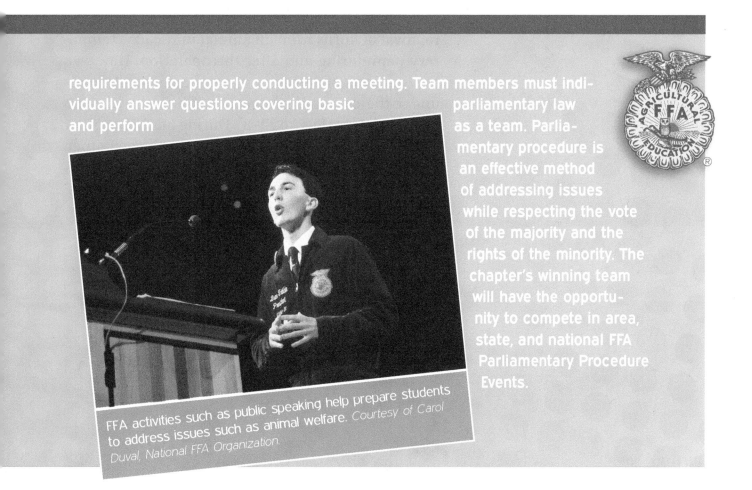

requirements for properly conducting a meeting. Team members must individually answer questions covering basic parliamentary law and perform as a team. Parliamentary procedure is an effective method of addressing issues while respecting the vote of the majority and the rights of the minority. The chapter's winning team will have the opportunity to compete in area, state, and national FFA Parliamentary Procedure Events.

FFA activities such as public speaking help prepare students to address issues such as animal welfare. *Courtesy of Carol Duval, National FFA Organization.*

FIGURE 20–7
Management operations are often viewed as being cruel.

castration, docking, and dehorning are cruel because they put the animal through a lot of pain (Figure 20–7).

Producers point out that these practices are done for the benefit of the animal. For example, the removal of horns from an animal may cause temporary pain during and after the operation. However, the purpose of horns is for self-defense. Animals under the care and protection of humans have no need for horns (Figure 20–8). During mating or other times, the animals may use the horns on each other. If they are removed at an early age, there is far less likelihood they will injure each other.

FIGURE 20–8
Domestic animals have no need for horns.

Castration will also prevent the animals from fighting. Neutering makes animals more docile and less aggressive. Normally, male animals fight to establish mating dominance. Castration prevents this behavior and prevents injury both to animals and humans. Practices such as docking (removing the tail) help keep the animals clean. Remember that all the practices used by producers are intended to help and not harm the animal. It is in the best interests of the producer and the animal to keep the animal healthy, comfortable, and contented.

Some practices, such as hot-iron branding, are used less now than in the past. Newer techniques, such as freeze branding and tattoos, are being used. These marking techniques cause less stress to the animals.

Several years ago, the National Cattlemen's Association adopted the following statement of principles on animal care, environmental stewardship, and food safety:

> I believe in the humane treatment of farm animals and in continued stewardship of all natural resources.
>
> I believe my cattle will be healthier and more productive when good husbandry practices are used.
>
> I believe that my and future generations will benefit from my ability to sustain and conserve natural resources.
>
> I will support research efforts directed toward more efficient production of a wholesome food supply.
>
> I believe it is my responsibility to produce a safe and wholesome product.
>
> I believe it is the purpose of food animals to serve mankind and it is the responsibility of all human beings to care for animals in their care.

Similarly, the National Pork Producers Council has adopted the following Pork Producers' Creed:

> I believe in the kind and humane treatment of farm animals and that the most efficient

production practices are those that are designed to provide comfort.

I believe my livestock operation will be more efficient and profitable if managed in a manner consistent with good husbandry practices as known and recommended by the animal husbandry community.

I believe in an open door policy to visitors to my farm, to all those who are sincerely interested in production methods and the welfare of animals, so long as they do not endanger the health and welfare of my animals and do not interrupt my production routine or impair the production process.

I believe in and will support research efforts designed to measure stress of farm animals and directed toward more efficient production of food and enhancement of the welfare of animals and man.

I believe it is the animal's purpose to serve man; it is man's responsibility to care for the animals in his charge. I will vigorously oppose any legislation or regulatory activity that states or implies interference with that responsibility.

Experimentation on Animals

Another issue of concern for animal welfare activists is the use of animals for research (Figure 20-9). For many years, animals have been used to test products and procedures for humans. For example, before any drug is tested on humans, it must be first tested on animals. Medical procedures such as operations and vaccinations are tested on animals. Most of the medical advances made by humans during the past century have come about through the use of animals to test treatments and medication. Research scientists argue that the use of experimentation on animals is well justified by these advances. They cite the

THE ETHICAL TREATMENT OF ANIMALS

FIGURE 20-9

Some people object to using animals for experimentation.

examples of diseases, such as polio, that have been almost eradicated through the use of research using animals in experiments.

Most people realize that research is carried out using mice, rabbits, and guinea pigs. However, during recent years much controversy has come about over the use of cats, dogs, and primates. Some people feel that it is worse to use animals that are "of a higher order." Those who oppose the use of animals in research contend that the knowledge gained through the research cannot justify the suffering the animals must undergo. They advocate the use of computer simulations to test treatments. Researchers point out the computer simulations are helpful, but nothing can ever replace real animals.

Research scientists have a much more difficult time justifying the use of animals to test such products as cosmetics. Activists insist that the suffering of animals should not be allowed merely to produce new products that are used only to make people appear more attractive.

Animal Rights Activists

Another group concerned with the treatment of animals is animal rights activists. They contend that animals have as much right to enjoy life and to be free as do humans. They believe that killing an animal is

FIGURE 20-10
Animals that have been bred to be cared for by humans could not live on their own.

just as wrong as killing a human. They oppose the raising of animals and the use of animal products. They also strongly oppose hunting and feel that animals should be free to live in the wild.

Opponents to animal rights activists contend that animals are not the equal of humans and that they exist for use by humans. They argue that if animals were turned loose, both animals and humans would suffer. Animals that have been bred to be cared for by humans could not care for themselves (Figure 20-10). If animals were turned loose without care, they would lead short, miserable lives. Domesticated animals live the best lives under the care of humans.

Opponents also point out that the loss of animal products would be a burden to humans. Nutrients obtained from the eating of meat would be difficult to replace from other sources. The absence of drugs and other pharmaceuticals made from animal by-products would create problems for those people whose lives depend on these products.

Animals left to live on their own would have a difficult time reproducing and surviving. When left entirely on their own in a closed ecological system, animal populations might be able to achieve a balance. However, human populations are large enough to prevent this from happening. A good example is that of deer populations that are allowed to grow entirely on their own. Unless the population is kept

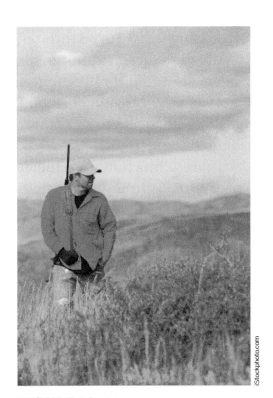

FIGURE 20-11
Hunting helps prevent wildlife from becoming overpopulated.

under control by measures such as hunting, there are too many of the animals (Figure 20-11). When this happens, the herds of animals starve or die of disease. A properly controlled population is healthier and the animal population can be sustained.

SUMMARY

In the future, agriculture will have to contend with an increasing number of people who advocate the ethical treatment of animals. This means that agriculturists will have to make sure that their management practices are proper and that they treat the animals well. In addition, they must make sure that the public is educated about how agricultural animals are treated. As with most issues, communication between the opposing sides is vital. This will help ensure that the ethically correct solution is found.

20 CHAPTER REVIEW

✻ Student Learning Activities

1. Some people feel that rodeos are cruel to animals. Research the topic and draw your own conclusion. List your reasons for making your conclusion. An alternative is to make a conclusion as to whether you feel that testing cosmetics on animals is justified. In class, debate others who hold the opposite opinion.

2. Make a list of all the ways animal producers protect their animals. Also make a list of management procedures that might appear to be cruel but are in the best interests of the animals. Be sure to include the reasons why they are beneficial.

3. Visit the local humane society and talk with the employees. With their help, devise a plan for your class to help stop the mistreatment of animals.

4. Talk to livestock show officials and determine what rules are in place to protect animals. Find out if there are other regulations that should be adopted.

True/False

1. _____ Today in the United States, only a small percentage of the people really understand the raising of farm animals.
2. _____ Most of the animals raised are produced on corporate-owned farms.
3. _____ Pigs are unsociable animals and like to be separated from other pigs.
4. _____ Most animals do better when raised with other animals of their own kind.
5. _____ Animals that are kept in confinement can receive better care than animals that roam free.
6. _____ Animals under the care and protection of humans have no need for horns.
7. _____ Castration keeps animals clean.
8. _____ Hot-iron branding is used less now than in the past.
9. _____ Animal rights activists believe that killing an animal is just as wrong as killing a human being.
10. _____ The National Pork Producers Council has adopted a Pork Producers' Creed.

Multiple Choice

1. Animal welfare activists disagree with producers over what constitutes proper
    a. neutering
    b. treatment
    c. confinement

2. Producers are in the business of raising animals for
    a. pets
    b. profits
    c. insect control

3. Animals that are placed in crates a week before they give birth are
    a. sows
    b. cows
    c. horses

4. A process performed to help keep animals clean is
    a. neutering
    b. beak trimming
    c. docking

5. An issue of great concern for animal welfare activists is the use of animals for
   a. food
   b. pets
   c. research

6. Before any drug is tested on humans, it must first be tested on
   a. insects
   b. animals
   c. bacteria

7. Animals that live the best lives under the care of humans are called
   a. confined
   b. domesticated
   c. wild

8. An example of a disease that has almost been eradicated through the use of research is
   a. polio
   b. cancer
   c. diabetes

9. Organisms that cause stress and discomfort to animals and may even shorten their lives are called
   a. herbivores
   b. parasites
   c. sows

10. A procedure that makes animals more docile and less aggressive is
    a. neutering
    b. dehorning
    c. confining

Discussion

1. What are two ways of thinking concerning the well-being of animals?
2. Why is it important that animals have pleasant surroundings?
3. Why is stress so detrimental to animals?
4. How do animals benefit from confinement?
5. What management practices do animal welfare activists disagree with?
6. How have scientists justified the use of animals for research?
7. Why do scientists have a more difficult time justifying the use of animals to test cosmetic products?

8. What are the beliefs of animal rights activists?
9. Explain how deer are a good example of the human impact on controlling animal populations.
10. What is the National Cattlemen's Association statement of principles based on?

CHAPTER 21
Selecting and Using Hand Tools

Student Objectives

When you have finished studying this chapter, you should be able to:
- Practice proper safety when using hand tools.
- Explain tool selection and care.
- Identify and properly use common woodworking hand tools.

Key Terms

hand tools
power tool
plane
customary measuring system
metric measuring system
square
measuring device
ruler
measuring tape
electronic measurers
marking tool
kerf
ripping
chisel
boring tool
brace
hammer
screwdriver
clamp
level
plumb bob

CHAPTER 21

Introduction

Throughout this book, you have learned about the fantastic world of agriculture. This dynamic industry represents much of the wealth of the United States. The tremendous amount of production came about as a result of research and development dealing with growing plants and animals. However, just as important as the developments in plant and animal science is the development of machinery and tools to aid in production.

As you know, agriculture began before recorded history. Thousands of years ago, humans had no tools but their bare hands. Because of their extraordinary reasoning ability, they began to devise tools that made their efforts more efficient. For example, a stone could be used as a hammer and a broken piece of flint could be used as a cutting instrument (Figure 21–1). As these crude tools were used, ideas for improvement surfaced. A stone could make a better hammer if it were shaped to fit the hand. Even more force could be applied if a stick were used for a handle. Broken flint could be shaped into a very effective knife or projectile point. Since that time, there has been an ongoing effort to make tools better.

From prehistoric time to our modern time, tools have been developed, enlarged, refined. Today, we have powerful machines, some of them two stories high, which are controlled by one person. These machines can do work that could not be done by hundreds of men without tools. Tools are multipliers of our strength, and are like millions of willing hands, working for our comfort. Without tools, there would be no civilization as we know it today.

Even through the development of complicated machine tools, **hand tools** have never lost their importance. They are still necessary in all

FIGURE 21–1
In prehistoric times humans made hand tools from stone.

FIGURE 21-2

A hand tool is any tool that is operated by hand to do work.

industries that involve production and maintenance, especially agriculture. A hand tool is defined as any tool operated by hand to do work (Figure 21-2). In contrast, a power tool is operated by some source of power other than human power. In this chapter, you will learn how to use hand tools, take care of them, and distinguish between the different kinds.

Simple Machines

From the most primitive of tools to our most complicated machine, almost all of them operate using one or a combination of six simple machines. These are the inclined plane, the wedge, wheel and axle,

FIGURE 21-3
An inclined plane is a slope that helps to move loads upward.

FIGURE 21-4
A screw is an example of an inclined plane.

screw, lever, and pulley. These simple machines are everywhere! Just look around you and you will find them.

An inclined plane is a slope that helps to move loads upward (Figure 21-3). Common examples are staircases and ladders. Both slope upward and make movement upward easier. Another example is the ramps that are placed at the end of a truck or trailer bed to load items such as lawn mowers or other heavy objects.

A wedge is a type of inclined plane. A wedge has a slope with a sharp edge at the end. It is usually used to separate material. An axe is a good example. The sharp edge is used to cut chips out of wood. Other types of wedges are chisels, punches, and splitting wedges.

Another type of inclined plane is called a screw (Figure 21-4). Everyone is familiar with screws that are used to hold wood together. If you look closely at the screw, you will see that the threads are really just a spiraling inclined plane. In fact, a screw can be considered to be a combination of an inclined plane and a wedge. As the screw is twisted, the inclined plane wedges into the wood and holds it tightly. Nuts, bolts, and jar lids are other examples of screws.

A lever is a long, rigid bar used to pry. It is used to generate more force for tasks such as lifting loads. A block placed under the lever is called a fulcrum. The best example of a lever and a fulcrum is the seesaw on the playground. Almost all tool handles are levers. Think about the handle of a hammer. You can hit much harder with the hammerhead on a handle than you can by using the hammerhead in your hand. The longer the hammer handle is, the harder the striking force. Shovels, picks, axes, hoes, and so on all make use of levers.

The wheel and axle consists of a round wheel that revolves around an axle. This simple machine is the basis for all carts, wagons, and automobiles. The principle is that the weight of an object is concentrated on the axle and the rotating movement of the wheel moves the weight. Variations of the wheel and axle are gears and pulleys (Figure 21–5). These devices are used to increase power.

Working With Hand Woodworking Tools

A common beginning point to learn about tools is using woodworking tools. Creating projects from wood can be fun. Not only do you have the enjoyment of learning to use tools, but you can build a useful project as well. In the next sections, you will learn about some of the basic woodworking tools.

Hand-Tool Safety

Before we discuss the various types of hand tools, it is important that you practice the proper safety procedures when using them. Hand tools, like any tools, can be very harmful if not used properly. Therefore, you should not just read but really learn the safety rules for any job you do and put the rules into practice (Figure 21–6).

There are certain steps that you should take when using a hand tool. They include:

- Wear safety glasses.
- Inspect tools before using. Do not use broken or damaged tools.
- Do not wear loose fitting clothing.
- Use the correct tool for the job. Do not use a tool that is either too big or too small for the job because it may result in injury or tool damage.
- Secure the work with a clamp, vise, or some other means of holding it. This prevents it from slipping and frees both hands to hold tools.

FIGURE 21–5
A pulley is an example of a variation of a wheel and axle.

FIGURE 21–6
When using hand tools safety must be practiced. For example, eye protection is a requirement.

- Grip tools firmly.
- Stand in a safe location. Neither you, nor anyone else, should stand directly in line with the tool's path of movement.
- Keep the work area clean and free of scraps and oil.
- Keep cutting tools sharp.
- Cut away from the body.
- Store tools properly. Make sure the sharp edges, if any, are down.
- Keep tools clean and free of excess oil and grease.

Importance of Proper Care of Tools

Knowing how to care for tools is very important, as their usefulness depends to a large degree on their care. Keeping tools well cleaned, oiled, and free from rust is essential, not only for ease in handling but also for lengthening their life. Tools should always be treated with oil when they are not in frequent use. Likewise, tools must be properly sharpened for good results (Figure 21–7). Tools that are dull and rusty will not work satisfactorily.

FIGURE 21–7
A sharp tool is much safer to operate than a dull one.

Cleaning Tools

Tools sometimes contact grease, paint, and other materials. These substances may damage tools and shorten their useful lives (Figure 21–8). You should always be careful with dirty tools, especially those that may have chemicals on them.

All grease on tools should be carefully removed with a soft material or, if necessary, by washing the tools with a safe cleaning solvent. Rust must be removed and prevented by

1. applying cleaning solvent and letting the tool sit for several hours;
2. rubbing and polishing with oil and pumice stone or with an emery cloth;
3. applying a thin film of light oil to the surface of the tool before it is put away.

FIGURE 21–8
Tools must be clean in order to operate efficiently. This person is cleaning a soldering iron. Dirty soldering irons do a poor job.

Measuring and Marking Tools

There is nothing more frustrating than cutting out a piece of lumber and discovering that it doesn't fit its intended opening. The problem usually results from the worker measuring either the lumber or the opening incorrectly. To prevent this, it is important to learn how to use measuring and marking tools.

Common measuring tools are squares, rulers, and electronic measurers. The United States uses two measuring systems: the customary measuring system and the metric measuring system (Figure 21–9).

- Customary—This measuring system is used in everyday life in the United States. Inches, feet, and yards are common measurements in the customary system. A measuring device using this system is graduated, or divided, into various fractions of an inch. The common graduations used are halves, quarters,

FIGURE 21-9
These rulers have both metric and customary scales.

eighths, and sixteenths of an inch. The customary system is widely used in agricultural mechanics.

* Metric—The metric system is the system used by all scientists and in everyday life in most parts of the world. It is based on a decimal system, with the meter being the base for measurement. Measuring devices in the metric system are marked as meters, decimeters, centimeters, and millimeters. The metric system is not as popular in the United States as in other countries.

Before using a measuring device, check to determine the system and graduations being used. Measurements in one system can be easily converted to the other.

Squares

A **square** is a tool used either to check whether a board is square at the end or to lay out a square line. Square means that a line or the end of a board is at a perfect 90-degree angle with the rest of the board. A square may be used to lay out and check other angles as well. Always mark a board with a square before sawing it. The common kinds of squares used in agriculture are framing squares, try squares, bevel squares, combination squares, and speed squares (Figure 21-10):

SELECTING AND USING HAND TOOLS

FIGURE 21-10
There are several different types of squares.

- Framing square—This type of square is in the shape of the large letter "L." The longer part of the square (24 inches) is called the body. The shorter part of the square (16 inches) is called the tongue. Framing squares are used for squaring boards and timbers, laying out rafters, determining if the end of the board is square, and squaring a board for sawing. Many different angles and patterns can be laid out using a framing square.

- Try square—The try square is smaller than the framing square and is usually from 4 to 12 inches in length. Only the blade of the try square has a measuring scale on it. The short part of the square is called the head. The primary use of this square is to determine whether a board is square. To protect it from being damaged and to ensure accurate measurements, the head is made of metal or wood framed in metal.

- Bevel square—These squares are adjustable so that they can be set to mark angles from 0 to 180 degrees. It is an excellent tool for marking angle cuts or checking the accuracy of angle cuts.

FIGURE 21-11
The speed square is relatively new and is helpful in laying out angles.

- Combination square—The combination square can be used for the same things as the framing and try squares but, because it has a movable head, its uses are more varied. It can be used as a straightedge, depth gauge, metric gauge, level, and marking gauge. Also, it can be used to lay out 45- and 90-degree angles.
- Speed square—A speed square is used for marking lumber and other materials for cutting. Information on the frame of the square often makes separate mathematical calculations unnecessary. Most speed squares have instructions for use. The use of this square is increasing in agricultural mechanics (Figure 21-11).

CAREER DEVELOPMENT EVENTS

If you enjoy working with tools, you will want to participate in the FFA Agricultural Mechanics Career Development Event (CDE). In this competition, you will work as a team with other students in your class. This event is centered on solving problems involving agricultural mechanics. Each of the problems in the CDE is a response to a problem that might be encountered in the workplace. According to the National FFA Organization, this CDE selects and awards those students and teams that demonstrate: (1) mastery of the subject matter and skills common to the topic; (2) effective communication skills; (3) superior problem-solving techniques; (4) an understanding of modern technology; and (5) the ability to function as team members working together and as individuals working alone.

The National FFA Agricultural Mechanics Career Development Event is divided into five areas:

1. Machinery and equipment systems
2. Industry and marketing systems
3. Energy systems
4. Structural systems
5. Environmental/natural resource systems

Measuring Devices

A **measuring device** is a tool used in making measurements. The device is normally marked in the standard units used in the measurement system. Several measuring devices are commonly used:

- Ruler—A **ruler** is a strip of wood, plastic, metal, or other material marked in increments and used for measuring and drawing lines. Rulers can be found in the customary measuring system, the metric system, or combinations of the two. The yardstick and meterstick are common examples. Most students are accustomed to using 12-inch rulers.

The Agricultural Mechanics Career Development event is centered on problem solving. *Delmar/Cengage Learning*

Individuals on each team work together and are evaluated as a team. The team is presented with a problem scenario, and members are provided with the materials and equipment needed to solve the problem. Teams have to organize, assign duties, and complete all of the tasks. Individuals are also scored and given individual awards based on points scored.

This career development event, like all of the career development events in the FFA, is designed to help you prepare job skills for the future. You learn to work as a team, communicate, and solve problems. While you are learning and developing skills, you get to travel and meet new people. At the same time, you will have loads of fun!

FIGURE 21-12
This is a zigzag, or folding, ruler.

FIGURE 21-13
Tape measures are convenient and easy to use.

- Zigzag ruler—This ruler is also known as the folding ruler (Figure 21–12). The zigzag ruler is made in sections that are easily folded for convenient use. They are commonly used in making measurements on lumber in building construction.

- Measuring tape—A **measuring tape** is a ruler of metal, plastic, or other material that is wound in a case. Graduations may be in the customary or metric system. Some automatically rewind, while others are rewound with a small handle. Tape measures are the most convenient, practical, and widely used measuring tool in the agricultural industry (Figure 21–13).

- Electronic measurers—**Electronic measurers,** also known as electronic tape measures, use ultrasonic sound waves to measure straight line distances. They are accurate for a distance of up to 45 feet. Conversions are made by the device from customary to metric system and vice versa.

Marking Tools

A **marking tool** is used to make marks on wood for cutting, placing, and measuring. When marking wood, the necessary tools are pencils or knives, marking gauges, and chalk lines (Figure 21–14):

- Pencil or knife—Use a sharp-pointed pencil or knife to mark a board for squaring. In fine work, a very fine line is necessary because a heavy line may cause a variation in the lengths of the pieces after they are sawed. This difference may be important when fitting small pieces together.

- Marking gauge—A marking gauge is often used when marking lines parallel to the edges of the material. However, this job may also be done with a straightedge or a pencil.

- Chalk line—A chalk line is used for marking a straight line on a board, wall, ceiling, or floor. It is a string that has been coated with colored chalk dust so that it leaves a line when snapped.

The steps in marking a piece of wood are:

1. square the piece with a square;
2. mark around the entire piece;
3. saw carefully by following the line.

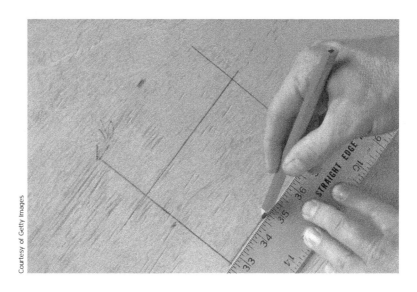

FIGURE 21–14
Pencils are used to mark wood at the proper place.

Saws

Handsaws are the most common cutting tools for wood used by agricultural workers. The handsaw is used to cut larger pieces of lumber into smaller pieces, according to the needs of specific jobs. The most common types used in agriculture are crosscut saws, ripsaws (Figure 21–15), compass saws, and coping saws.

Saws are generally categorized according to the number of points or teeth they have per inch of blade. An 8-point saw, for example, has 8 teeth for every inch of blade, whereas a 10-point saw has 10 teeth for every inch. The more points per inch, the finer or smoother the cut made by the saw.

A cut made by a saw is called a kerf. To start the kerf with the saw, it is important for the blade to be on the side of the line that will become the scrap piece of lumber. Then the saw is drawn backward in a smooth stroke. Long, smooth forward and backward strokes are then made, following the line drawn. The

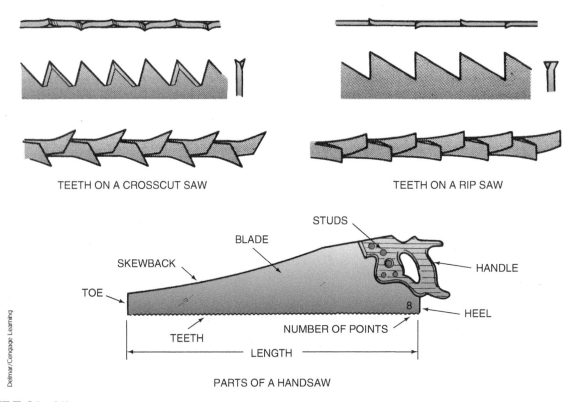

FIGURE 21–15
Hand saws are used for crosscutting or ripping, depending on the shape and angle of the teeth.

FIGURE 21-16
A coping saw is used for cutting curves or other irregular cuts.

blade must always be on the side of the waste lumber. The last few strokes are made very slowly and with slightly less pressure. This helps prevent the board from splitting as the cut is completed.

- Crosscut saws—As the name indicates, these saws are used for cutting across the wood grain. A 10-point crosscut saw is the most commonly used in the agricultural industry. However, crosscut saws are also available in 9 and 12 points. The crosscut saw is held at a 45-degree angle to the wood being cut, and the other hand is used to steady the material.
- Ripsaws—Ripping a board means to cut it lengthwise along the grain to make strips of narrower width. The ripsaw is used for this work. Its teeth are coarse, with 5½ points per inch being the most common. The teeth chisel the wood rather than cut it as the crosscut saw does. Because of this difference, the saws are sharpened differently. The ripsaw works best if held at a 60-degree angle to the material being cut.
- Compass saw—These saws are used for cutting curves, circles, and irregular shapes in pieces of lumber. Before a circle or other shape can be cut in wood, it is necessary to drill a hole large enough for the end of the blade to penetrate and start the kerf.
- Coping saw—This saw is also used for cutting curves or making other irregular cuts (Figure 21–16). The coping saw is different than the compass saw in that its blade is thin and removable and it has a great many more teeth along the blade so that it can make a much finer cut. Cuts with the coping saw should be made on down strokes only.

Holding Lumber in Place for Sawing

Place a long piece of lumber on sawhorses or other supports while sawing it. Stand so that you can put one knee on the piece being sawed. Place a short piece of wood in a vice or on a bench hook. For a very long piece of lumber, use two bench hooks, one at each end of the piece.

Wood Chisels

A **chisel** is a wedge-shaped cutting tool. It is designed to make special cuts in wood, such as grooves or notches. Chisels are also used to shape and trim wood (Figure 21–17).

The chisel consists of a blade and handle. The cutting edge of the blade is beveled, or angled. The opposite end of the blade is called the tang. This end fits into the handle. The two types of chisels frequently used in carpentry work are tang chisels and socket chisels.

The tang chisel is one in which the tang extends into a wooden or plastic handle. In a socket chisel, the handle is fitted into a socket formed by the metal in the chisel.

Chisels are used by workers by pushing the chisel forward, with the bevel (angle) down and the chisel positioned at an angle to the wood. The chisel can also be driven forward by tapping the handle or socket with a mallet or hammer. The degree of the angle usually determines how much wood is to be chiseled in one forward motion. If the chisel is held at a high angle, more wood will be cut away.

FIGURE 21–17

A wood chisel is used to trim and shape wood.

Boring Tools

A **boring tool** is a tool used to make holes or change the size or shape of holes. Boring tools include bits, drills, reams, and the devices used to turn them. The most common boring tool is the brace and auger bit (Figure 21-18).

A **brace** is the device for holding and turning an auger bit. Braces can either be the ordinary type or the ratchet type. An ordinary brace is satisfactory for all general work where there is plenty of room for making complete revolutions with the handle. A ratchet brace is constructed so that complete revolutions of the handle are not necessary. Consequently, it is convenient for use in corners or other places with little room to work.

The hand drill and electric drill are other tools used for boring holes in wood. Bits for the electric drill operate at high speeds and are different from bits used in the hand drill and the brace.

FIGURE 21-18
A brace and bit is used to bore holes in wood.

Hammers

A **hammer** is a tool made of durable metal with a handle made of wood, metal, or other material. There are two kinds of nailing hammers: one has a curved claw and the other has a straight claw. They are often referred to as curved-claw and ripping-claw hammers, respectively. The curved claw is particularly valuable for pulling out nails. The ripping claw is best used to rip apart or tear down wooden structures and to pry two boards apart (Figure 21-19).

Hammers have two kinds of striking faces. One is flat, and the other is slightly rounded. The slightly rounded head, called a bell-faced hammer, is used primarily to drive a nail slightly below the surface of the wood without leaving hammer marks.

Hammers are also classified by the kind of handle. Hammers with wooden handles are lighter and used in framing and general carpentry. Those with steel handles are used for heavier work. Hammers

FIGURE 21-19
A ripping-claw hammer (above) is used to rip apart nailed boards. A curved-claw hammer is used to pull nails. Both are also used to drive nails.

with steel handles and straight claws are used for ripping. Hammers with fiberglass handles are used for general work, depending on the claw shape.

SUPERVISED AGRICULTURAL EXPERIENCE

If you have an interest in tools and hardware, you could benefit from completing an SAE at a hardware store. SAEs are designed to provide hands-on experience for students to develop their skills in agricultural career areas. To begin, you must first be enrolled in an agriculture class. Your teacher will help you to design an SAE that combines your interests and skills learned in the classroom, shop, and laboratory. By becoming active in an SAE, you will develop skills in your area and learn how to use those skills on the job. You will also learn how to earn, save, and invest money.

An SAE in a hardware store is a type of placement program. Placement programs involve placing students on farms, in agricultural businesses, in school labs, or in community facilities to provide a "learning by doing" environment. This experience is done outside of normal classroom hours and may be paid or nonpaid. A placement SAE in a hardware store requires supervision by agriculture education teachers and employers, in cooperation with parents. These people work with the students and assist them in their development and achievement of their educational and career goals.

Along with the responsibility of maintaining an SAE comes record keeping. Record keeping is one of the most important components of an SAE. Students maintain a list of skills learned during the time spent in the classroom and at work. Material inventories, income, expenses, and accomplishments should be kept along with managerial decisions, labor, and advancements made on the job. With a long-term SAE, good record keeping, and FFA involvement, a student can earn well-deserved recognition. Proficiency awards for superior SAEs are among the many honors earned through agricultural education.

Screwdrivers

A **screwdriver** is a tool with a handle on a long metal shank. Screwdrivers come in all shapes and sizes and are designed to fit the heads of a variety of screws.

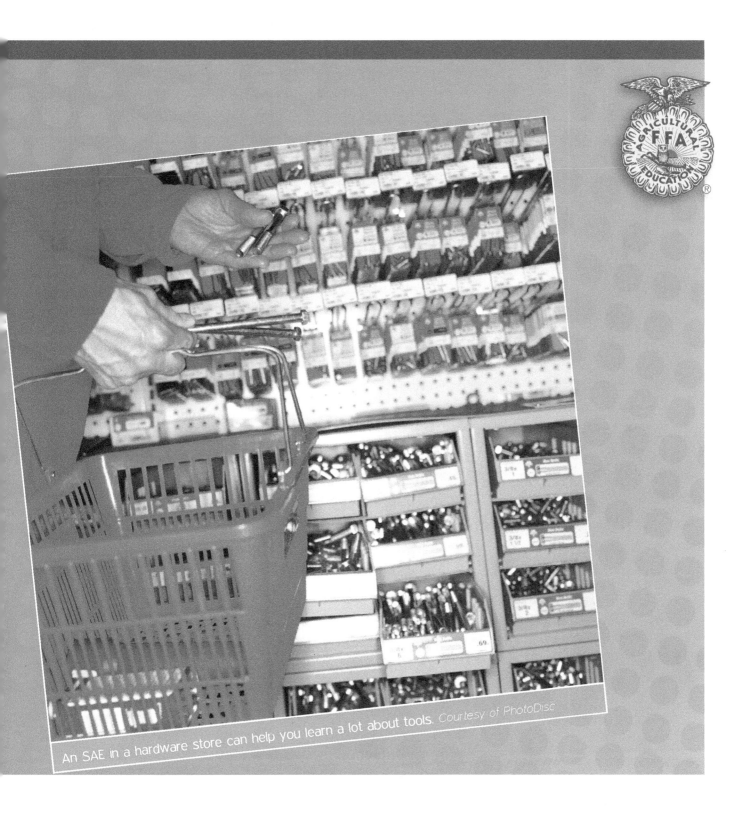

An SAE in a hardware store can help you learn a lot about tools. *Courtesy of PhotoDisc*

FIGURE 21-20
The screwdriver on the left is a Phillips and the one on the right is a standard.

Whenever materials are to be joined together with screws, and that happens in nearly all agricultural career fields, the screwdriver is needed to drive the screws.

There are basically two kinds of screwdrivers for woodworking, the standard and the Phillips (Figure 21-20). The standard head is flat and tapered. The Phillips screwdriver has a cross-shaped bit.

The size of the screwdriver used depends on the head of the screws to be driven. The tip of the screwdriver should not be wider than the screw. If it is wider, it will scar the wood around the head of the screw. The tip of the screwdriver should not be larger or smaller than the slot in the screw. If it is larger, it will not extend to the bottom of the slot in the screw; if it is smaller, it will slip out of the slot easily.

No tool is perhaps abused more than the screwdriver. It should not be used other than to drive screws. Excessive force exerted on the long handle of a screwdriver can snap the handle. This could in turn cause hand or bodily injury.

Clamps

A **clamp** is a device for holding pieces of wood or metal in position until they are fastened together by nails, screws, bolts, glue, or welding (Figure 21-21). Clamps may also be used to exert pressure so that materials may be fitted together properly. Common types of clamps are hand screw clamps, "C" clamps, bar clamps, and miter clamps.

Levels

A **level** is used to determine whether a surface is flat (horizontal). Levels involve using a bubble in a freeze-resistant liquid in a small tube-shaped glass container. Markings on the tube are used to assess the position of the bubble and determine if the surface is level. It can also be used to determine if an object such as a fence post or wall is plumb. Plumb means that it is perfectly vertical (straight

FIGURE 21-21
Clamps are used to hold wood together for gluing or nailing.

FIGURE 21-22
A spirit level uses a bubble enclosed in glass to indicate when an object is level or plumb.

up and down). The bubble is entirely between the two markings if the surface is level or plumb (Figure 21-22).

Levels come in various lengths, ranging from a few inches to over 4 feet. They are made of wood or metal. Metal levels are often preferred because they do not warp. In selecting a level, choose the longest one that is convenient for the job. A longer level is more accurate in determining the surface position.

Plumb Bob

A plumb bob, also known as a plummet, is used to determine a plumb, or vertical, line (Figure 21-23). A plumb line is at a right angle to a level horizontal line. This is important in building construction to ensure that the walls, floors, and ceilings fit properly. A plumb bob may be used to find a point directly beneath an overhead point. A breeze may destroy the accuracy when a plumb bob is being used, by moving it around.

FIGURE 21-23
A plumb bob is used indicate a vertical line.

SUMMARY

Humans have used hand tools since prehistoric times. Tools help us to work more efficiently and also to do work we could not do with our bare hands. Today there is a wide variety of different hand tools to choose from. Learning to identify and properly use hand tools will allow us to build, maintain, and repair our homes in an effective way. Safety should always be practiced when using any type of tool. Tools used safely and properly will give many years of service.

CHAPTER REVIEW

Student Learning Activities

1. Closely examine a bicycle. Make a list of all the simple machines that make up the bicycle.
2. Use the Internet to explore kinds and sources of hand tools. Several Web sites are listed below. The Hand Tool Mall has excellent safety information.

   Matco Tools—http://www.matcotools.com

   The Stanley Works—http://www.stanleyworks.com

   Hand Tool Mall of the Internet—http://www.toolsource.com

   Hand Tools Institute—http://www.hti.org

   Old Forge Tools—http://www.oldforge.com
3. Prepare a bulletin board with pictures of the various hand tools described in this chapter and captions with descriptions of their uses.

True/False

1. _____ You should wear loose fitting clothing when working with hand tools.
2. _____ Tools that are cleaned, oiled, and free from rust aid in ease of handling.
3. _____ Tools play a small role in agriculture.
4. _____ The problems that occur in measuring and marking are usually the fault of the person who did the measuring.
5. _____ All handsaws are made the same.
6. _____ The cutting edge of a chisel is an example of a wedge.
7. _____ Braces are used to bore holes in wood.
8. _____ Two types of hammers are the curved claw and the straight claw.
9. _____ A hammer is needed to drive screws in agriculture.
10. _____ A plumb bob is also known as a plummet.

Multiple Choice

1. A weighted string used to find a level vertical line is called a
   a. level
   b. plumb bob
   c. square

2. Levels make use of a glass tube filled with
   a. water
   b. salt
   c. a freeze resistant liquid

3. Standard and Phillips are types of
   a. screwdrivers
   b. saws
   c. hammers

4. Hammer handles can be made of
   a. wood
   b. fiberglass
   c. wood or fiberglass

5. A device used to turn an auger bit is called a
   a. brace
   b. clamp
   c. twister

6. Ripsaws should be held at an angle of
   a. 40 degrees
   b. 80 degrees
   c. 60 degrees

7. Electronic measurers use _____ to measure straight line distances.
   a. ultrasonic sound waves
   b. microwaves
   c. radio waves

8. A square that is used for laying out rafters is called a
   a. try square
   b. combination square
   c. framing square

9. Square refers to an angle of
   a. 90 degrees
   b. 45 degrees
   c. 60 degrees

10. A measuring system that uses meters, centimeters, and millimeters is called the
    a. standard system
    b. customary system
    c. metric system

Discussion

1. Why is the proper care of tools important?
2. Name the different types of simple machines.
3. What are the common kinds of squares?
4. What is the difference between a ripsaw and a crosscut saw?
5. What is a kerf?
6. What tools are commonly used for marking?
7. What is a level used for?
8. What tool is perhaps abused most frequently?
9. How can you classify hammers?
10. What are plumb bobs used for?

CHAPTER 22
Small Engine Operation

Student Objectives

When you have finished studying this chapter, you should be able to:

- Discuss the uses of small engines.
- Observe all safety precautions when working with small engines.
- Distinguish between two- and four-cycle engines.
- Explain how a four-cycle engine operates.
- Explain how a two-cycle engine operates.
- Identify all the systems in a small engine.
- Explain how the different systems of a small engine work.

Key Terms

internal combustion engine
engine
motor
combustion
engine timing
cycle
stroke
four-stroke cycle engine
two-stroke cycle engine
top dead center (TDC)
bottom dead center (BDC)
intake stroke
compression stroke
compression ratio
power stroke
exhaust stroke
scavenging
magneto ignition system

Perhaps the greatest impact on agricultural production was the invention of the **internal combustion engine.** This type of engine burns fuel inside the engine. Until its invention, engines burned fuel outside the engines. These engines usually produced steam for a power source. We generally think of internal combustion engines as running large tractors and harvesting equipment. However, a huge number of small engines are in use. These are usually one cylinder engines that are portable and run small equipment.

Today, there are over 80 million small air-cooled engines in use in the United States (Figure 22–1). These small engines are a popular power source for many different types of equipment.

One reason why small engines are so widely used is because the engines are easily adaptable to many jobs and conditions. They are easily adaptable because

1. the engines require no outside source of power, therefore, they are compact and relatively lightweight;
2. they are cooled by air instead of liquid coolant;
3. the engines are relatively easy to service and repair, which allows operators to do their own maintenance.

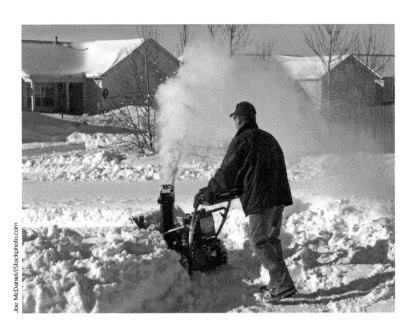

FIGURE 22–1
There are over 80 million small engines in use in the United States.

Examples of equipment powered by small engines are: lawn mowers, chain saws, go-carts, minibikes, tillers, pumps, and sprayers.

Small Engine Safety

Before the operation and mechanics of small engines are discussed, it is important to address safety. Annually in the United States, 150,000 people are injured while mowing grass. Over 55,000 are treated in hospitals. There are 55,000 toes and 18,700 fingers amputated by lawnmower blades. Only 9 percent of these injuries are caused by mechanical failure, whereas 91 percent are caused by human error.

Some of the common ways people get hurt by small engines include touching a power mower's swirling blades with a hand or foot, being struck by a flying object hurled by rotary mower blades, catching a body part while operating or repairing a machine while the power is on, having a machine overturn, falling off a machine, and receiving an electrical shock.

In order to prevent minor or severe injuries while operating small engines, it is important that certain safety precautions be taken (Figure 22–2). Operators should adhere to the following accident prevention reminders:

1. Dress for the job. Wear safety glasses and leather shoes.
2. Keep hands and feet a safe distance from all moving parts.
3. Do not allow burning material to get near oil, solvents, or gasoline.
4. Stop the engine and allow it to cool before refueling.
5. Whenever possible, handle gasoline outdoors.
6. Disconnect the spark plug cable when making any adjustments on machines, to prevent the possibility of accidental engine startup.

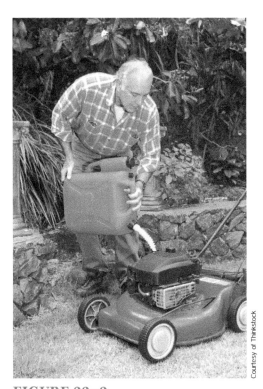

FIGURE 22–2
Saftey should be foremost in operating small engines. Never fill a fuel tank when the engine is hot.

7. Never disengage any of the safety equipment or devices.
8. Keep all shields in place.
9. Do not operate engines above the specified speed.
10. Do not overload engines or force equipment beyond the designated capacity.
11. Never start and run an engine inside a building in which the ventilation is inadequate.
12. Never use gasoline as a solvent or to clean engine parts.
13. Dispose of gasoline properly by pouring it into an approved container. Never pour it onto the ground or down a drain.

Importance of Proper Care of Small Engines

Small engines are rugged machines that are built to last. Most small engines are designed to operate continuously at or near top operating speeds to prevent overload. Small engines are designed and built to operate under full load at top operating speeds for at least 1,000 hours without requiring major repairs, if the engine is properly serviced and maintained. Modern small engines take little time and effort to maintain (Figure 22–3).

The most important aspects are keeping the oil at the correct level and keeping the air breather clean. Before you operate or attempt maintenance on a small engine, read and thoroughly understand the owner's manual.

Operating Principles of Small Engines

A person who intends to service or repair a small engine must have at least a working knowledge of the operating principles of the engine. Before you can troubleshoot problems or fix the engine, you have to understand how the engine operates. You need to understand the operation of the various parts of the

FIGURE 22-3
Modern small engines take little time and effort to maintain.

SMALL ENGINE OPERATION

FIGURE 22-4
An engine produces power using energy generated within the device. A small engine is a good example.

engine and be able to identify each step in the operation cycle. Once you know this, locating and fixing problems will be much easier.

Power Production

An engine is a device or machine that can produce power on its own—independent from an external source of energy (Figure 22-4). An engine has an energy source (fuel) and converts the fuel to usable power. The power produced is usually in the form of mechanical power.

How is an engine different from a motor? Generally, a motor is considered to be a device or machine that produces power from energy supplied by an external source. For example, an electric motor takes the electrical power produced somewhere else, such as a generator, and then converts it to mechanical power (Figure 22-5).

Combustion

Combustion is the chemical process of oxidation that produces heat. The combustion process requires three things: fuel, air (oxygen), and heat. An internal combustion engine uses the process of rapid combustion to operate. Sufficient heat must be produced from the combustion process to produce usable power.

FIGURE 22-5
A motor produces power using an outside source. An electric motor is an example.

Internal Combustion Engines

Internal means that the combustion occurs in an enclosed chamber (called the combustion chamber). The heat produced is converted to mechanical power. An internal combustion engine is a machine or device that is capable of converting heat energy into mechanical power. The mechanical power produced is usually in the form of a rotating force.

SUPERVISED AGRICULTURAL EXPERIENCE

One of the parts of a high school Agricultural Education program is the Supervised Agricultural Experience (SAE). This part of the program is designed so that students can put into practice what they learn in the classroom. These activities are planned, practical activities conducted outside the classroom. They may be conducted at the student's home or may be organized at a business such as a small engine repair shop. The experience is called supervised for a good reason. The agriculture teacher visits with the student at the site of the SAE to give advice and help direct the experience. SAE programs are explained in more detail in Chapter 25.

The National FFA Organization provides awards for students who excel in their SAEs. Each award is called a proficiency award, and they all are based on the goals achieved by the student and the skills and concepts learned. They are also based on the records kept by the student. The Agricultural Mechanics Design and Fabrication Proficiency Award is one of three agricultural mechanics proficiency awards available to FFA members. The others are (1) Agricultural Mechanics Repair and Maintenance and (2) Agricultural Mechanics Energy Systems. These proficiency awards are presented to individual FFA members who have effectively produced and marketed their talent in the agricultural mechanics industry. The student must have a Supervised Agricultural Experience (SAE) in a selected special agricultural mechanics area. All of the agricultural mechanics proficiency award areas are in the entrepreneurship/placement category. This means that the student must be the entrepreneur of his or her own agricultural mechanics project or work in the agricultural mechanics production area.

The small engine, like any gasoline-burning engine, is an internal combustion engine. Four events must take place within the engine for an internal combustion engine to run continuously. The four events and the order in which they occur are:

1. Intake—the process of getting the fuel and air required for combustion into the combustion chamber.
2. Compression—the process of compressing the air-fuel mixture in the combustion chamber.

Planning and goal-setting for winning proficiency awards should begin as soon as a student enters the Agricultural Education program. The SAE can last until the student has been out of high school for a year. Talk with your teacher and begin planning your SAE and setting goals to win a proficiency award.

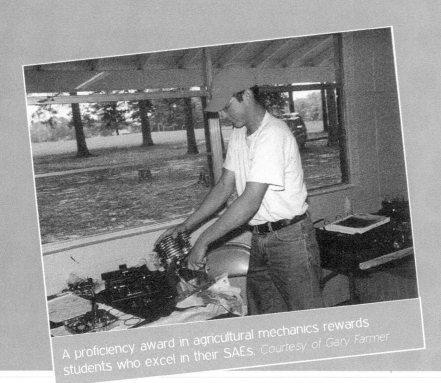

A proficiency award in agricultural mechanics rewards students who excel in their SAEs. *Courtesy of Gary Farmer*

3. **Power**—the ignition, burning, and expansion of the air-fuel mixture.
4. **Exhaust**—removing spent products of combustion from the combustion chamber.

These four events make up an internal combustion engine cycle. They must occur in the proper sequence and at the proper time. Engine timing refers to the proper timing and sequence of these events. Once a cycle is completed, another cycle immediately begins.

Terms Used to Describe the Operation of Engines

There are four terms used to describe the operation of small engines. They are:

1. **Cycle**—the completion of the four events: intake, compression, power, exhaust.
2. **Stroke**—the full travel of the piston in one direction, either toward the crankshaft or away from the crankshaft.
3. **Four-stroke cycle engine** (four-cycle engine)—an engine requiring four strokes of the piston to complete one cycle. One stroke of the piston is required for each of the four events.
4. **Two-stroke cycle engine** (two-cycle engine)—an engine requiring only two strokes of the piston to complete a cycle of operation.

The Four-Stroke Cycle Engine

The four-stroke cycle engine (Figure 22–6) operates on a series of four strokes, or piston movements, to a cycle. The piston moves up and down within a cylinder that is only slightly larger than the piston diameter.

The cylinder is inside a mass of metal (usually aluminum) called a block. At the top of the block are valves that let in fuel and let out exhaust fumes. The piston is connected to the crankshaft that converts the piston's up-and-down motion to a rotary motion,

SMALL ENGINE OPERATION

FIGURE 22-6
The main parts of a 4-cycle engine. Note that the valves on this one are in the head. Some small engine valves are in the block.

much like a person pedaling a bicycle. The up-and-down movement of the person's legs is converted to a rotary movement of the bicycle wheels.

As the pistons move up and down, the movement in one direction is called a stroke. When the piston reaches the top of its stroke, the volume of the combustion chamber is at its smallest. This position is called top dead center (TDC). When the piston reaches the bottom of the stroke, the cylinder volume is at its greatest. This position is called bottom dead center (BDC). Each of these strokes serves a particular purpose (Figure 22-7).

Intake Stroke

The first stroke of the cycle is the intake stroke. This occurs when the piston moves down, the intake valve opens, and the mixture of fuel and air enters the cylinder. At the completion of this stroke, the piston is at the bottom of the cylinder and both valves are closed.

Compression Stroke

The second stroke is the compression stroke. This occurs when the piston reaches the bottom of the cylinder on the intake stroke and starts upward.

FIGURE 22–7
This diagram shows all of the strokes of a four-cycle engine. A complete four-stroke cycle is completed 25 times every second.

Gasoline engines compress an air-fuel mixture. At the top of this stroke, the air-fuel mixture has been compressed into the area in the head called the combustion chamber. The degree that the air-fuel mix is compressed is referred to as the **compression ratio.** This is calculated by comparing the volume of the cylinder at TDC to the volume of the cylinder at BDC. The volume of a cylinder is calculated by the formula $V = \pi r^2 h$, where $\pi = 3.14$, r = the radius of the cylinder, and h = the height of the cylinder. If the diameter of the piston is 2½ inches, the radius would be 1¼ inches (radius is one-half the diameter). If the height of the cylinder at TDC is .5 inches, the volume at TDC would be 1.53 cubic inches. If the height of the cylinder at BDC is 4 inches, the volume at BDC would be 12.25. The volume of the cylinder at BDC (12.25) divided by the volume at TDC (1.53) is 8.00, or an 8:1 compression ratio.

Power Stroke

As the piston nears the top of the cylinder on the compression stroke, a spark from the spark plug ignites the air-fuel mixture. The spark is precisely timed by the ignition system to occur at exactly the correct time. The resulting rapid expansion forces the piston down, creating the power stroke. During the power stroke, the intake and exhaust valves must remain tightly sealed. The larger the cylinder and piston, the more power the engine can produce. This is because there is more room for the air-fuel mixture and, the larger the room, the greater the amount of energy that is released on ignition of the gases.

Exhaust Stroke

The fourth and final stroke of the cycle is the exhaust stroke. This occurs when the piston reaches the bottom and starts upward. The purpose of the exhaust stroke is to remove the burned air-fuel mixture so fresh air and fuel can be brought into the combustion chamber. During this stroke, the exhaust valve opens and, as the piston moves up, the exhaust gases are forced out through the exhaust system. The process of moving the exhaust gases is called scavenging. For an engine to operate efficiently, the exhaust must be scavenged as completely as possible.

At the end of the exhaust stroke, the piston is at TDC, and a new engine cycle begins. It is important to keep in mind that a complete cycle of the four-stroke cycle engine happens in a very short period of time. For example, if a single cylinder engine is operating at a speed of 3,000 revolutions per minute (rpm), the engine is operating at 50 revolutions per second. Therefore, the engine is completing 25 cycles per second and each piston stroke is taking $\frac{1}{100}$ of a second to complete.

The Two-Stroke Cycle Engine

The two-stroke cycle engine is also widely used in the agricultural industry. This type of engine completes the intake, compression, power, and exhaust stages in two strokes. It is also a piston- or reciprocating-type internal combustion engine. Generally, the operation of the two-stroke cycle engine is more difficult to understand, because more than one engine cycle is occurring at the same time. Two-stroke cycle engines are commonly used in agriculture for small engine applications because they are more lightweight and can be operated upside down. Chain saws, weed eaters, leaf blowers, and other small power tools are examples of implements that use two-cycle engines (Figure 22–8). These engines require that oil be mixed with the gasoline. Instead of having a lubricating system that contains engine oil, the oil is simply mixed with the gasoline.

The two-stroke cycle engine (Figure 22–9) operates on the principle of using the movement of both ends of the piston to create compression and a partial vacuum. The top of the piston is used to operate and create the combustion chamber in the cylinder. The bottom of the piston is used to draw the air-fuel mixture into the engine through an opening in the cylinder wall. This opening is called a port. The port is opened and closed by a thin metal mechanism

FIGURE 22–8
Two-cycle engines are lightweight and can be operated in any position. Note the safety precautions of the operator. She is wearing hearing protection, eye protection, and is safely dressed.

FIGURE 22–9
The strokes of the two-cycle engine are illustrated here.

called a reed valve. This valve is opened and closed by vacuum and pressure created by the piston moving up and down.

Intake and Exhaust Strokes

As the piston nears the bottom of its stroke, it uncovers the intake and exhaust port(s). Because the air-fuel mixture in the crankcase is under pressure, it rushes through a passage to the intake port and enters the cylinder. This incoming gas mixture pushes air, or exhaust, out of the cylinder. Therefore, intake and exhaust functions occur with very little movement of the piston.

Compression Stroke

At the end of the intake/exhaust stroke, the cylinder is filled with an air-fuel mixture. The piston moves upward, closes the intake and exhausts ports, and compresses the air-fuel mixture trapped in the cylinder. At the same time, a new supply of air and fuel rushes into the crankcase.

Power Stroke

The spark plug fires to ignite the mixture at or near TDC. The burning and expanding gases drive the piston downward through the power stroke. This same downward movement puts pressure on the new air-fuel mixture in the crankcase. Thus, the engine completes its cycle of intake, compression, power, and exhaust with only two strokes of the piston.

Comparing Four- and Two-Stroke Cycle Engines

The four-stroke and two-stroke cycle internal combustion engines have specific characteristics that allow both to be used for various agricultural purposes.

Generally, the four-stroke cycle engine operates more quietly, is heavier, has a longer life, and is cleaner burning than a two-stroke cycle engine. However,

Two-Stroke Cycle Engines	Four-Stroke Cycle Engines
Lighter weight	Heavier weight
Operates in many positions	Operates in limited positions
Higher power-to-weight ratio	Lower power-to-weight ratio
Engine oil usually mixed with fuel	Engine oil in a reservoir
Louder operation	Quieter operation
Higher engine speeds	Slower engine speeds
More vibration	Smoother operation
Rough idling operation	Smoother idling operation

the two-stroke cycle engine is lighter, can be operated in a wider range of positions, has fewer moving parts, and can be used more for smaller jobs.

Diesel Engines

Many small engines are diesel engines. In the past, diesel engines have been mostly used for large equipment and trucks. However, in recent years, diesel engines have been made that are smaller and lighter than the older heavy diesels. Small diesel engines are used to operate little tractors, generators, and irrigation pumps.

These engines differ from gasoline engines in that they have a different ignition system. Recall that the fuel mixture in a gasoline engine is ignited by a carefully timed spark from a spark plug. Diesel engines are ignited by the heat of compression. When air is tightly compressed, it becomes hot. Rapidly pump a bicycle pump and feel the bottom of the pump. The heat you feel is the heat of compression. In a diesel engine, an injector is used instead of a spark plug. A high pressure stream of diesel fuel is injected into the cylinder at the end of the compression stroke. The heated air causes the fuel to ignite and the power stroke begins. The compression ratio for a diesel is much higher than that of a gasoline engine in order to get enough heat to ignite the fuel.

The advantages of diesel engines are that they are usually tougher, produce more power, and last longer than gasoline engines. In addition, fuel costs are generally lower. The disadvantages of diesel engines are that they are heavier, they are noisy, and they often have an unpleasant odor. Despite these disadvantages, small diesel engines are becoming more popular.

Engine Systems

An engine operates through the coordination of several different systems that must function properly for the engine to operate. These systems are much like the systems of the body. Our digestive, skeletal, circulatory, and nervous systems must all function properly before the body can operate efficiently. All systems of the body rarely get sick at the same time. Likewise, when an engine malfunctions, it is usually only a part of one system that malfunctions. That part, however, can cause the entire engine to malfunction.

The Compression System

The primary purpose of this system is to compress the air to increase the energy resulting from the combustion of the fuel. The compression system of the internal combustion engine uses pistons that move up and down in a cylinder. The piston and cylinder must form a leak-proof combustion chamber for the engine to operate. The airtightness of the combustion chamber is a major design feature of internal combustion engines. If the compression system loses compression, the engine will not operate properly. To ensure that the fit is tight, the pistons have grooves all the way around the top of the

FIGURE 22-10
Rings set in grooves on the piston expand against the cylinder wall to create a tight fit.

piston. Inside these grooves are rings of steel or cast iron that expand against the cylinder wall and create a tight fit (Figure 22-10).

The Intake System

The purpose of the air intake system is to allow clean air to efficiently enter the combustion chamber. Air is composed of approximately 20 to 25 percent oxygen. Oxygen is required for combustion. The air intake system cleans dirt particles from the air without causing a significant restriction to the free flow of the air, which would decrease engine power output.

Fuel System

The fuel system is designed to deliver clean fuel to the combustion chamber and only allow the amount of fuel necessary for efficient operation. The fuel needs to be sent at the correct time and in the best proportion with the air for the engine to operate properly. There are a variety of fuel systems and components to meet the needs of different kinds of engines, fuels,

FIGURE 22-11
Most small engines have a device called a carburetor that mixes gasoline with air in the proper ratio.

and applications. However, most fuel systems operate based on the same basic principles. Small engines have a device called a carburetor that mixes the gasoline with air (Figure 22-11).

Suction from the intake stroke sends the fuel mixture into the combustion chamber.

Exhaust System

The exhaust system performs the task of removing the burned gases from the combustion chamber. The exhaust system also helps with engine noise reduction and heat transfer. A device called a muffler is attached to the cylinder. The exhaust fumes pass through the muffler. This process "muffles" or deadens a lot of the sound from the engine. In recent years, the exhaust system has been designed to help control engine emissions that may damage the environment.

Cooling System

The cooling system is designed to manage the heat produced by the combustion of the air and fuel. Every engine has an operating temperature that is best for that engine. The task of the cooling system is

FIGURE 22-12

Fins on the engine block and head help dissipate heat from the engine.

to allow the engine to reach the correct temperature and then maintain this temperature under varying conditions.

Almost all small engines are cooled by the air. Most large engines are cooled by a system that circulates liquid coolant through the engine. In air-cooled systems, the heat from the engine components is transferred directly to the surrounding air. This process is aided by fins that are built into the cylinder block (Figure 22-12). The fins provide a larger surface area that comes into contact with the cooling air.

Ignition System

The ignition system is designed to ignite the air and fuel mixture in the combustion chamber. Spark ignition engines (gasoline) use an electrical spark to ignite the air and fuel mixture. These systems create an electrical spark in the combustion chamber at the end of the compression stroke. With correct ignition timing, the ignition spark occurs at the precise moment to ensure the most complete and efficient combustion of the air-fuel mixture. The electrical spark is generated from a magnetic field created by magnets. This system is called a **magneto ignition system**. A device called a coil increases the amount of voltage (electrical pressure) generated by the magneto. The current flowing through the spark plug can be around 40,000 volts.

SMALL ENGINE OPERATION

FIGURE 22-13
Current flowing through the spark plug tries to jump across the gap at the end of the spark plug. This creates a spark.

This current flows through the spark plug and tries to jump across the gap at the end of the spark plug. This creates a spark (Figure 22-13).

Lubrication System

Internal combustion engines need lubricants to keep engine parts operating smoothly. The engine uses a lubrication system to keep the engine parts lubricated to reduce friction, remove heat, keep engine parts clean, and provide tight seals on various engine parts to keep dirt out and compression pressure in. The lubrication system is designed to maintain the optimum amount of engine oil lubricant on various parts. Even small engines have lubrication systems, although they may not be as complex as the systems used on larger engines. It is extremely important to keep the engine filled to the proper level with clean oil. This means that the oil must be checked before the engine is started. Oil should be changed at intervals suggested by the manufacturer of the engine.

Starting System

All internal combustion engines require some type of system to start. Some small engines use a manual operation system, such as a recoil rope starter, which the operator pulls; an example of this type of system is a go-cart. Most large engines use an electrical starting system, which uses an electrical motor to start the engine.

SUMMARY

Small engines are an important part of the agricultural industry. Although they exist in a variety of types, they all operate basically the same way. They may seem complicated, but they are made up of several less-complicated systems. An understanding of the basic principles of operation and how the systems work together is essential to the operation, maintenance, and repair of the engines.

CHAPTER REVIEW

Student Learning Activities

1. Disassemble a small engine and label the parts. Reassemble the engine.
2. Create a list of the various small engines on a farm, school facility, or agribusiness. Determine whether the engines are two- or four-cycle engines and whether they are gasoline or diesel engines.

True/False

1. _____ Small engines require an outside source of power.
2. _____ Nine percent of small-engine injuries are caused by the operator.

3. _____ An engine can produce power on its own.
4. _____ Internal combustion means that combustion occurs in an enclosed chamber.
5. _____ The four terms used to describe the operation of small engines are cycle, stroke, combustion, and four-stroke cycle engine.
6. _____ Two-stroke cycle engines are lighter weight than four-stroke cycle engines.
7. _____ The exhaust stroke of a four-stroke cycle engine occurs when the piston reaches the bottom and starts upward.
8. _____ The two-stroke cycle diesel engine is common in agriculture.
9. _____ Lubricants are used to keep engine parts running smoothly.
10. _____ Most large engines use an electrical starting system.

Multiple Choice

1. How many hours can small engines operate without requiring major repairs?
   a. 500
   b. 1,000
   c. 1,500
   d. 2,000

2. What does the combustion process not require?
   a. fire
   b. fuel
   c. air
   d. oxygen

3. A(n) _____ is the full travel of the piston in one direction, either toward or away from the crankshaft.
   a. cycle
   b. intake
   c. stroke
   d. rotation

4. When the piston reaches the top of its stroke, it is said to be at _____.
   a. the top
   b. dead center
   c. top dead center
   d. bottom dead center

5. What is the purpose of the exhaust stroke?
   a. remove the unwanted gas
   b. remove the fumes in the cylinder
   c. discard the excess air
   d. removed the burned air-fuel mixture

6. Which of the following is an example of a piece of equipment that operates using a two-stroke cycle engine?
   a. lawn mower
   b. go-cart
   c. leaf blower
   d. mini-bike

7. The purpose of the _____ system is to allow clean air to enter the combustion chamber.
   a. intake
   b. fuel
   c. compression
   d. ignition

8. What does a lubrication system not do?
   a. remove heat
   b. reduce friction
   c. reduce speed
   d. keep parts clean

9. Most large engines use a(n) _____ starting system.
   a. gasoline
   b. electrical
   c. spark ignition
   d. key-driven

10. Annually, in the United States, _____ people are injured mowing grass.
    a. 50,000
    b. 100,000
    c. 200,000
    d. 150,000

Discussion

1. List the 13 safety precautions that should be taken when operating small engines.

2. What are the four events that must take place for an internal combustion engine to run smoothly?

3. Describe the difference between TDC and BDC.
4. What is the main difference between a two-stroke cycle engine and a four-stroke cycle engine?
5. List five characteristics of four-stroke cycle engines as compared to two-stroke cycle engines.
6. What is the difference in the fuel sources of gasoline and diesel engines?
7. What are the eight engine systems?
8. List and describe the two types of cooling systems.
9. What is the purpose of a lubrication system?
10. Why do you think proper small engine operation is important in agriculture?

CHAPTER 23
Biofuels

Student Objectives

When you have finished studying this chapter, you should be able to:

- Describe the importance of energy.
- Discuss the need to find alternatives to fossil fuels.
- Name at least four types of biofuels.
- Describe the process used in making ethanol.
- Explain why biodiesel is a promising alternative diesel fuel.
- Discuss some of the concerns about using biofuels.

Key Terms

fossil fuels
nonrenewable energy
biofuels
ethanol
enzyme
mash
fermentation
distillation
by-product
stillage
E85
E10
biodiesel
catalyst
B20
biomass
renewable energy
carbon cycle
manure
methane
natural gas
lagoon

Almost all aspects of our lives demand that we use some form of energy. This energy may be in the form of electricity used to light and heat our homes, run our electrical appliances, or power our iPods. Energy may also come in the form of oil used to heat our homes or heat water for our use (Figure 23-1). A vast amount of energy is used to power our cars, trucks, trains, and heavy equipment. Just think of all the things you contact daily that need energy. The clothes you wear took a lot of energy to produce. The fiber had to be grown or produced in a factory. The fiber had to be woven into cloth, then made into the shirt, blouse, pants, or dress. The desk you sit at also took a lot of energy to produce. Even your body uses energy in the form of the food you eat each day. The food you eat required the use of energy for its production and transportation to the grocery store. All of these items required transportation; that took a lot of gasoline, diesel fuel, and oil. The point is that we depend on energy to maintain our lifestyle. Without energy it would be difficult to continue life as we know it.

A very large part of our energy needs come from what we call **fossil fuels**. The two most obvious examples are coal and petroleum. They are called fossil fuels because they originated from plants and animals that existed millions of year ago. These forms of energy

FIGURE 23-1
Energy to heat our homes may come from electricity or heating oil.

are also called nonrenewable energy resources. This means that when the supply is depleted, the supply is gone forever.

Many people feel we are in an energy crisis. An energy crisis occurs when the demand for energy such as petroleum-based fuel exceeds the amount of fuel produced. As large countries like the United States continue to grow and smaller nations are beginning to develop larger economies and industries, the world demand for petroleum will continue to increase. The United States uses around 21 million barrels of petroleum per day. The next time you are in a large city at rush hour, consider all the fuel being consumed by all the vehicles you see. Also consider that you are observing only one city on one day. Think of all the many large cities in the world that burn that much fuel in a day (Figure 23–2). We are using more petroleum per day than we can obtain from wells in our country. Much of our petroleum is exported from the Middle East. In addition, we get oil from places such as Nigeria, Mexico, and countries in South America. Some of the countries we import oil from are considered to be unfriendly to the United States.

Another problem with petroleum-based fuel is that it is created from nonrenewable sources. Petroleum is a fossil fuel that was created through the decomposition of organic materials that began

FIGURE 23-2

A tremendous amount of petroleum is consumed each day.

millions of years ago. The high demand for fossil fuels such as petroleum means that sources are being rapidly depleted and one day fossil fuel sources will disappear. Scientists have long been aware of this problem and are working to find a solution. Biofuels offer one such solution. **Biofuels** are liquid, solid, or gaseous energy sources created from organic materials that come from renewable sources such as agricultural crops. Biofuels are sometimes called agrofuels. Many scientists are hopeful that biofuels will not only help solve fuel supply problems but will also help cut down on harmful emissions such as greenhouse gases. Created from virtually any carbon-based material, the sources of biofuels are renewable. The most common forms of biofuel are ethanol, biodiesel, biomass, and methane.

Ethanol

Ethanol is a form of alcohol that is considered to be one of the most promising alternative fuel sources. Ethanol is seen as a great alternative fuel option because it is clean burning and easily produced. Ethanol is created through the fermentation and distillation of grains. You may have read stories about people who used to make and sell illegal whisky. These producers were called moonshiners and the product they made was ethanol (Figure 23–3). Today's manufacturers

FIGURE 23–3

Moonshiners used to make ethanol in stills such as this.

use the same basic principles as the old moonshiners. The production is much more sophisticated and is on a much larger scale (Figure 23-4). Basically any grain can be used to produce ethanol, but the most widely used is corn.

There are two main manufacturing processes used in producing ethanol. One process is called dry milling and the other is called wet milling. In dry milling the whole corn kernel is ground into meal and the meal is mixed with water. Enzymes are then added to the corn meal and water mixture. An **enzyme** is a substance that causes a chemical reaction. This reaction causes the meal and water to convert to a form of sugar called dextrose. This slurry of material is called **mash**. Next, the mash is moved into a high temperature cooker where bacteria are killed. Once the mixture is cooled, yeast is added and the fermentation process begins. **Fermentation** is a process in which simple sugars are broken down into cellular energy. The result of the fermentation process is ethanol and carbon dioxide. After 40 to 50 hours, the mixture is placed in distillation columns where the liquid ethanol is separated from the solid mash mixture. **Distillation** is the process in which a liquid is heated until it converts to a vapor, then cooled until it converts to a liquid again. This purifies the liquid. Once the ethanol has been distilled through two separated

FIGURE 23-4
Ethanol is made in large plants such as this.

processes, the result is a liquid that is almost 100 percent ethanol. The ethanol is then ready to go to suppliers and retailers.

The by-products of the dry milling process are also put to use. A by-product is a product other than the main or intended product. In this example, ethanol is the main or intended product and the material left, called stillage, is the by-product. Stillage is used in the production of a high quality livestock feed. Another by-product is the captured carbon dioxide gas. It is used for many different things such as providing the carbonation for soft drinks.

Wet milling produces many different by-products. The wet milling process begins by steeping the grain in water and a weak sulfurous acid for one to two days. This results in the separation of the corn kernel into many parts. The slurry of corn kernels is next separated into its various parts. The small part of the corn kernel from which the new plant emerges, if the kernel is planted, is called the germ (Figure 23–5). The corn germ is pressed into corn oil that is used for cooking and other purposes. The water and acid mixture used to steep the corn is dehydrated, or dried, and sold to create livestock feed. It can also be used as a product called Ice Ban that is used as an environmentally friendly way to deice roadways. Remember from Chapter 10 that grains contain a sticky substance

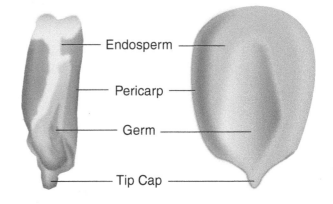

FIGURE 23-5

The corn germ is the part of the corn kernel from which the new plant emerges.

called gluten. The corn gluten is filtered and dried, resulting in a product used by the poultry industry for feed. The leftover starch and water from the mash is either used to produce modified cornstarch, corn syrup, or is distilled into ethanol in much the same process used in dry milling.

During the 1970s leaders from the oil-producing countries severely limited the amount of petroleum exported. This caused prices to rise to extremely high levels. The Brazilian economy was hit hard because at that time the country imported more than 80 percent of its fuel supply from foreign sources. This energy crisis caused the Brazilian government to realize that the country must become more energy self-sufficient, and they began mixing ethanol with their gasoline. Today ethanol represents about 30 percent of the country's fuel use, and most of the automobiles manufactured in Brazil have the ability to run on either gasoline or ethanol. Most of the Brazilian ethanol is produced from large crops of sugarcane (Figure 23–6).

In the past few years, the United States has increased its ethanol production in order to offer an alternative to petroleum-based fuels. In the United States corn, sorghum, and wheat are the most commonly used grains for the production of ethanol. By 2012, the United States Energy Department projects that 30 percent of the country's corn crop

FIGURE 23-6
In Brazil ethanol is made from sugarcane.

FIGURE 23-7
E85 fuel is a combination of 85 percent ethanol and 15 percent gasoline.

will be used for the production of ethanol. This is likely to expand even more as fuels such as E85 are used. E85 is a fuel made from 85 percent ethanol and 15 percent gasoline, and is mainly sold in Midwest (Figure 23-7). The U.S. Congress has declared that E10, a mixture of 10 percent ethanol and 90 percent gasoline, be the required fuel offered at fueling stations. Almost all new cars and small trucks are now made to run on this mixture.

Biodiesel

Biodiesel is another alternative fuel source that is receiving a lot of recognition today. Like ethanol, biodiesel has already been widely used across the world for some time (Figure 23-8). Biodiesel is made by mixing oils from plant and/or animal sources with petroleum-based diesel fuel to provide a way to make petroleum fuel last longer. Biodiesel is made by using fats or lipids that come from substances such as vegetable oils or animal fats that have been refined for use as diesel fuel. A good source for these fats is the large amounts of plant- and animal-based oils used to fry food. Each day the fast food industry fries vast amounts of French fries, chicken, fish, and other foods. Fresh oil must be used to make the food tasty. The old oil can be filtered, refined, and added to diesel fuel to make biodiesel (Figure 23-9).

FIGURE 23-8
Biodiesel has been used across the world for some time.

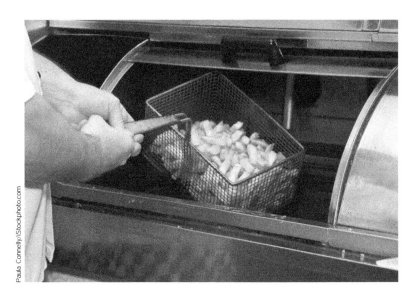

FIGURE 23-9
Oil from the fast food industry can be filtered, refined, and used as biodiesel.

The diesel engine has been around since 1898. A German refrigerator engineer named Rudolph Diesel set out to prove that an engine could be created that would ignite fuel without a spark. After many tries he created the diesel engine, which ignites fuel using heat generated by a great amount of air compression in the engine cylinders. Modern tractors, buses, large trucks, bulldozers, and train locomotives all use diesel engines (Figure 23-10). In fact, many cars now have diesel engines. Diesel engines use fuel more efficiently than gasoline engines. Also, diesel is less expensive to make than

FIGURE 23-10
Diesel engines are more efficient than gasoline engines. They are used in large transfer trucks.

gasoline. This has led the automobile industry and consumers to begin a renewed appreciation for the diesel engine. The creation of biodiesel has further pushed this movement toward more diesel cars.

The use of vegetable oil as fuel in diesel engines is not a new idea. Rudolph Diesel experimented with many different types of fuel sources for his new engine, including powdered coal and peanut oil. He finally realized that peanut oil was the best fuel and designed the engines to run on it.

Today in the United States the most commonly used sources for biodiesel are soybean oil and recycled frying oil from the restaurant industry. There are many ways to produce biodiesel but they are all very similar. The process includes transforming the fats in the oil into a fatty acid and then into compounds called esters that are then combined with alcohols such as methanol. Catalyzation is the process in which the fats in the oils are turned into fatty acids. Catalyzation is the speeding up of a chemical reaction through the addition of a catalyst. A catalyst is a substance that is added to a mixture in order to accelerate a chemical reaction without destroying the added substance itself. Heat is an example of a catalyst.

Biodiesel has become popular over the past several years because of the ease of making diesel engines compatible with biodiesel. Many modern pickup

FIGURE 23-11
Diesel engines are now used in cars and pickups. This truck has a diesel engine.

trucks and cars now use diesel engines (Figure 23–11). A car or truck with a diesel engine can be converted to run off vegetable oil. People who drive these converted cars report that their vehicles run just as smoothly as before, with the same, if not improved, fuel mileage. There are both positive and negative aspects to using straight vegetable oil. One of the positive aspects is that cars running on vegetable oil give off very few harmful emissions when compared to cars running on petroleum-based fuels. Another is that vegetable oil is a renewable fuel source. The most common problems associated with using straight vegetable oil are that it can clog the fuel lines if the system is not flushed properly and the oil filter must be changed quite often. In addition, people report that their automobiles give off a slight odor of whatever food was last cooked in the oil.

Vegetable oils that have been processed into biodiesel do not have these complications. Biodiesel does not usually require any modifications to a standard diesel engine. Many city and state governments now use biodiesel as a way of powering vehicles such as buses. A common mixture is called B20, a mixture of 20 percent biodiesel and 80 percent petroleum based diesel. European and South American countries are also seeing an increase in the consumption of biodiesel. In 2003, the European Union passed a directive that requires 2 percent of the fuels used for transportation to come from renewable sources such as biodiesel. Germany leads the European Union in the amount of biodiesel consumed each year. In Germany most biodiesel is obtained from rapeseed.

Biomass

Biomass is another example of a **renewable energy** source that has promise as an alternative to petroleum-based fuels. Biomass is a term for large amounts of organic waste products such as lawn clippings, wood chips, or manure (Figure 23–12). Some biomass is created through the large-scale cultivation of crops such as sorghum, corn, and sugarcane. Although coal and petroleum are organic fuel sources, they are not considered biomass. In order for something to be considered biomass, it cannot be affected by geological change.

Another reason that coal and petroleum are not considered biomass is because they have been out of the carbon cycle for too long. Biomass is part of the carbon cycle. The **carbon cycle** is an important function of the earth, where plants use carbon dioxide and sunlight to produce energy and grow in a process called photosynthesis. Plants later die and in the case of biomass the dead plant life is used as

FIGURE 23-12

Waste wood chips such as this can be used for making biofuel.

fuel, releasing carbon dioxide into the atmosphere. Carbon dioxide acts as a greenhouse gas that traps heat and keeps the Earth from freezing. Carbon dioxide is also given off by the burning of fossil fuels. Some scientists say that too much carbon dioxide is being released into the air by this burning and is causing climate change.

Wood is the most common form of biomass. It is often burned to heat homes and supply power to businesses (Figure 23–13). Businesses such as paper manufacturers and lumber mills use the scrap wood that is the by-product of manufacturing processes as fuel. Like wood, manure is another by-product that is quickly becoming popular as a fuel source. Many livestock yards around the United States are beginning to harness the energy of their herds' waste. Dried manure has long been used as a fuel source. If the manure has less than 20 percent water content, it can be burned directly. In the nineteenth-century United States, pioneers headed west during Manifest Destiny would pick up dried manure along the wagon trails when wood was scarce, in order to keep their campfires burning at night. Today, researchers at Texas A&M University are working to determine the best ways to use dried manure on a large scale. In Hereford, Texas, a town known as the cattle capital of the world, manure is being used to power an

FIGURE 23-13

Wood is the most commonly used biomass for fuel. Many people heat their homes with wood.

ethanol plant. This is a great example of an industry making use of its waste products to its benefit. The producers in the livestock business benefit from the sale of their stock's manure, and the ethanol producers save money by not having to spend as much on petroleum based fuel. As more research is conducted in the area of using biomass as biofuel, more business and domestic uses will be found for our everyday waste items.

Biomass can also be converted to ethanol. The process is complicated but it holds promise for the future. In the logging industry a lot of biomass is left from tree limbs and other debris. It is thought that this large amount of biomass can one day be a cheap source of material to make ethanol.

Methane

Methane is the largest component of natural gas and is another type of biofuel. **Natural gas** is a form of gas created by the transformation of plant and animal life into fossil fuel. Many homes in the United States are heated by piped-in natural gas. It is clean burning and therefore desirable as a fuel source. One of the largest problems with the use of methane gas is that it is very difficult to transport. It is also highly flammable, meaning that it can explode easily, which makes a natural gas leak very dangerous (Figure 23–14).

FIGURE 23-14
Natural gas is transported through large pipes. This man is checking for leaks.

FIGURE 23-15
Manure from dairy operations is washed into a storage pond called a lagoon.

The livestock industry, in particular the dairy industry, is using methane to provide power for their farms. Most dairy farms use a lagoon to store the manure produced by their herds. A lagoon is a large lined pool where liquefied manure is stored in order to keep it from seeping into the ground and contaminating groundwater. Generally the manure stored in the lagoon is used to fertilize fields, but more recently dairy farmers have begun to use it to power their operations (Figure 23–15). The conversion of manure to electricity is fairly simple. First, the liquefied manure is piped into another storing facility, like a silo. Once in the silo, the manure is heated to around 100 degrees Fahrenheit, the point at which the natural bacteria present in the manure begin to create methane gas. As the bacteria begin to break down the manure, the methane gas escapes to the top of the liquid, where it is collected and stored. The methane is then used to power a special generator that produces electricity.

Concerns

Biofuels can be one solution to the growing global problem concerning the future of our energy sources. As with any solution there are both advantages and disadvantages to the use and production of biofuels. Each of the different types of biofuel discussed in this chapter presents both promises and concerns.

FIGURE 23-16
A concern is that making ethanol raises the price livestock producers pay for feed.

Ethanol is an alternative to petroleum fuels, and because it is already being mixed with gasoline in the United States, there has been a lot of discussion regarding its use. One of the advantages of ethanol is that it comes from a renewable source. However, there are serious concerns that the United States will be unable to produce enough ethanol from grain to serve as a serious alternative to gasoline. Another concern with the use of ethanol is the rising price of corn. Because ethanol production is increasing, the demand for corn is increasing as well, causing the price of the crop to rise. This presents a problem to the livestock industry, which depends on corn for feed, because suddenly the cost of feed has risen (Figure 23-16). Other critics say that we cannot produce enough ethanol from corn or biomass to make a substantial difference in our energy supply. They also point out that with current technology, it costs more energy to produce ethanol than the energy ethanol produces. In addition to this, some people report lower gas mileage when using ethanol. There are reports of ethanol being corrosive to a vehicle's engine block because of the high water content of the fuel.

Many of the concerns regarding biofuels are about the dependence on crops for our fuel source. If our goal is to find a more stable fuel source because of the problems in petroleum production, we face similar issues when we rely on crops for our fuel. Crops are vulnerable to climate, drought, disease, and insects, making the crop yield each year unpredictable. People are also concerned that once crops begin going toward the production of biofuels, we will not have enough crops for our food supply.

SUMMARY

All of our modern economy runs on energy. Most of the necessities and luxuries in our lives depend on a reliable source of energy. For many decades the majority of our energy needs has come from fossil fuels. There can be no doubt that eventually we will run out of fossil fuel as an energy source. When that happens, new sources of energy must be found. While there may seem to be some disadvantages to biofuels at the moment, science and research are continually working to find new ways to make biofuels work to our advantage. With new research and developments, biofuels can become a major source of energy. When this technology is fully developed, agricultural producers and consumers will both benefit.

23 CHAPTER REVIEW

Student Learning Activities

1. Go to the Internet and research alternative forms of energy other than biofuels. Examples might include hydrogen or nuclear energy. Make a list of the advantages and disadvantages of each energy source.
2. Create a list of material you see in your local area that might be used as a source of biomass.
3. Choose another country and research the sources of energy used in that country. Present a report to the class on your findings.

True/False

1. _____ Most of the energy we use is from nonrenewable sources.
2. _____ Fossil fuel is a renewable resource.
3. _____ The total petroleum need for United States is about five thousand barrels a day.
4. _____ Much of the petroleum we use comes from the Middle East.
5. _____ Ethanol is a fuel made from corn.
6. _____ By-products from making ethanol are discarded.
7. _____ Biodiesel is a fuel composed of vegetable or animal oil combined with gasoline.
8. _____ Diesel engines will run on pure vegetable oil.

9. _____ Manure can be used as a source of fuel.

10. _____ Biomass is another term for grain used in making fuel.

Multiple Choice

1. Energy may come in the form of
   a. electricity
   b. fuel oil
   c. food
   d. all of the above

2. Coal and petroleum are examples of
   a. biomass
   b. renewable energy
   c. fossil fuels
   d. carbon dioxide

3. We import petroleum from
   a. Nigeria
   b. Mexico
   c. the Middle East
   d. all of the above

4. Ethanol is a form of
   a. gasoline
   b. alcohol
   c. natural gas
   d. diesel fuel

5. A country that has made good use of ethanol as a fuel is
   a. Brazil
   b. Colombia
   c. Nigeria
   d. Canada

6. Diesel engines were originally designed to run on
   a. coal dust
   b. gasoline
   c. ethanol
   d. peanut oil

7. The burning of fossil fuels gives off
   a. carbon dioxide
   b. oxygen
   c. methane
   d. biomass

8. Wood is a form of
    a. biomass
    b. gluten
    c. stillage
    d. natural gas

9. Manure can be used to make
    a. biomass
    b. biodiesel
    c. methane
    d. ethanol

10. A lagoon is used to store
    a. biomass
    b. by-products
    c. natural gas
    d. manure

Discussion

1. What is the difference between a renewable and a nonrenewable energy source?
2. Briefly explain why we need to develop biofuels.
3. Name the most common forms of biofuel.
4. Name two methods used to produce ethanol from corn.
5. What is meant by fermentation?
6. Why has biodiesel become popular?
7. List some of the common sources of biomass.
8. Explain how manure can be used as a fuel.
9. What is the source of natural gas?
10. Explain two concerns about the use of biofuels.

CHAPTER 24
Biotechnology: The Future of Agriculture

Student Objectives

When you have finished studying this chapter, you should be able to:
- Relate why agriculture will always exist.
- Cite examples of how biotechnology has been used in the past.
- Explain the current uses of genetic engineering.
- Discuss some future uses of genetic engineering.
- Explain the concerns over genetic engineering.
- Discuss careers in agricultural science.

Key Terms

biotechnology
artificial insemination
embryo transfer
clone
genetic engineering

Over the years many industries and occupations have come and gone. As jobs outlive their usefulness, the industry dies out. One industry that can never die out is agriculture. People must have food, shelter, and clothing. All of these needs, especially food, depend on agriculture.

Even though agriculture will always exist, the way things are done in agriculture will change. Just look at all of the changes that have occurred since humans first began to till the earth using sharpened sticks. Although changes took place over thousands of years, most have occurred during the last century (Figure 24–1). Futurists tell us that the changes you will see during your lifetime will be greater than any that have occurred so far. That is hard to imagine!

Biotechnology

The most profound changes in the future will occur in biotechnology. This term refers to the application of biological and engineering techniques to living organisms. This concept is not new. In fact, biotechnology has been in use for thousands of years. Perhaps the earliest biotechnology was used to make cheese and wine. Grape juice and milk spoiled very rapidly. Through the use of certain types of bacteria, the milk

FIGURE 24–1
Since ancient times many changes have occurred in the way humans grow food. This machine is used for planting Brussels sprouts.

could be fermented to form cheese and the juice could be fermented to form wine (Figure 24–2). The use of these bacterial agents transformed highly spoilable foods into foods that could be kept for a long period of time. In the same manner, food such as fresh cabbage could be fermented to make sauerkraut, which would last longer.

Ancient biotechnology was not limited to food preservation. Legend has it that hundreds of years ago, Arabs first used artificial insemination. This process involves collecting sperm from the male and placing it in the female reproductive tract. The Arabs have always been noted for the beautiful horses they grew. According to the legend, Arabs from one tribe would sneak into the camp of enemy tribes and collect semen from the prized stallions of the enemy chieftains. They would then use the semen to impregnate their mares. This artificial mating produced fine horses.

Reproductive Technology

Reproductive technology has progressed dramatically since the time when the Arabs first began artificial insemination. Both artificial insemination and embryo transfer are now commonplace technologies. Embryo transplant (sometimes called

FIGURE 24-2

Long ago grape juice would spoil unless it was fermented into wine.

embryo transfer) involves collecting embryos from a superior female animal. The embryos are then transplanted into females of lesser quality. These animals then carry the embryos to birth. Both dams and sires are chosen through the use of computers. The ability to freeze and store semen and embryos has tremendously affected the production of superior animals (Figure 24-3). With these techniques superior animals can reproduce at a much more rapid rate.

Scientists can now even **clone** animals. Cloning is the ability to produce an organism that is genetically identical to another. Plant scientists have been able to clone plants for some time using a technique called tissue culture. Animals have been more difficult to clone. However, several companies now offer frozen embryos for sale that have been cloned from superior parents.

Another reproductive technique that is beginning to be used is the development of all-male or all-female embryos. This technology is valuable to producers who only produce one or the other sex. For example, if a producer is in the business of hatching pullets to be grown into laying hens, only the females

FIGURE 24-3
The use of artificial insemination and embryo transfer has greatly advanced the animal industry. This technician is placing semen (sperm) into the reproductive tract of a cow.

are valuable. Now, about half of the eggs that hatch are male. If a process could be developed to produce only eggs that would hatch female chicks, the efficiency of the hatchery would be doubled. Another example is the production of heifers for dairies. Bull calves are not nearly as valuable as female calves that will one day produce milk (Figure 24-4). The technology already exists to produce embryos that are only male or only female. The challenge is to make the process economically feasible.

Genetic Engineering

The newest form of biotechnology is genetic engineering. Genetic engineering is the altering of the genetic makeup of organisms. Genes are the substances that determine the characteristics of an individual organism. Genes are made of a substance called deoxyribonucleic acid, more commonly referred to as DNA. The arrangement of DNA on genes determines the characteristics of an organism. Specific segments of DNA on the gene control specific characteristics. This is true whether the organism is a simple plant, an earthworm, or a human being. This is why scientists are able to identify the parentage of animals.

Great progress is made when scientists are able to completely map the DNA and genetic makeup of an organism. One of the first uses of this new

FIGURE 24-4
Dairy producers place a much greater value on female calves than they do on male calves.

technology was the manufacture of human insulin. This substance is produced by the human body and helps to regulate the metabolism of carbohydrates. If the body does not manufacture enough insulin, a condition known as diabetes results. Previously, insulin to treat diabetes was harvested from the pancreases of slaughtered cattle and pigs. Because only a tiny amount could be obtained from an animal, the process was expensive. Also, cattle insulin was not exactly the same as human insulin and some people were allergic to it.

This problem was solved through the use of genetic engineering. Scientists were able to find the gene part, or DNA, that regulates the production of insulin. They then spliced this segment into the gene

AGRISCIENCE RECOGNITION

The FFA Agriscience Student Recognition Program not only rewards the students, but it is designed to educate parents, school personnel, and the public about the career opportunities available in agriscience. The program recognizes students who are pursuing challenging school-based courses focused on the scientific principles and new technologies of agriculture. The winners of this award receive financial assistance to attend college. This helps to produce a reliable supply of agriscience graduates to meet society's future needs.

To be eligible for participation in the Agriscience Recognition Program, the student must be an FFA member who is a high school junior or senior, or a first-year student in college. The student must be taking academically demanding courses, including classes in agriculture, and be planning a career in agriculture that requires at least a bachelor's degree. The student must submit an application and list of agriscience activities and completed projects. These activities will be judged by awarding points for skill, scientific thought, thoroughness, clarity, and creative ability. The student's academic achievement and school and community activities are also judged.

The student's agriscience project must follow the scientific research process. Each phase of the project must be identified, explained, and recorded. Tables, lists, or graphs are a good way to illustrate observations made during the experiment. The agriscience project should fall into a plant science, animal science, natural of the *E. coli* bacteria. The bacteria that produce insulin multiply and pass the capability along to the next generation. Because the genetically altered bacteria are weaker than the ordinary bacteria, they have to be grown under carefully controlled conditions.

When the proper number of bacteria are reproduced, they are taken apart to retrieve the insulin produced by the bacteria. The insulin is purified, and the remains of the bacteria are destroyed. The genetically altered bacteria produce a ready, relatively inexpensive supply of insulin.

Bacteria have become the manufacturing centers for many substances that have made the lives of humans better and more productive. Vaccines that are used to impart immunity to animals and humans

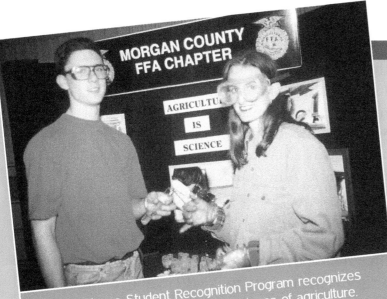

The Agriscience Student Recognition Program recognizes students who have excelled in the science of agriculture. Courtesy of Blane Marable, Morgan Co. High School, Madison, GA.

resources, earth science, energy production and manangement, food science, or engineering science category. Applications should include letters of recommendation from an employer, instructor, counselor, or other school official, plus pictures showing the student participating in the project.

The Agriscience Student Recognition Program application must be submitted through the agriculture instructor. The state winner will be presented a scholarship for use at the college of the winner's choice and is eligible to compete for regional and national awards. Regional and national winners can win scholarship money that totals over $8,000.

are produced in much the same way as insulin. Hormones that control growth and other bodily functions are also produced using genetic engineering.

Through genetic engineering, scientists have developed plants that are resistant to herbicides. Herbicides are chemicals that are used to kill weeds or unwanted plants. When the weeds are growing among a crop, a herbicide must be used that will kill the weeds and not harm the crop. Genetic engineering can develop crop plants that are tolerant of herbicides. This is a tricky process because the genetically manipulated crop cannot be guaranteed to retain all of the characteristics of the nonmanipulated crop. For example, grain that is grown for food still has to produce well and also must contain the proper amounts of nutrients.

Scientists have developed the ability to determine exactly where the part of the gene is located that controls certain characteristics. They can do this for some characteristics on some organisms. Just think of the possibilities when they are able to map out and identify exactly where every characteristic is located for every organism. For example, some farm animals, such as sheep and cattle, are ruminants. Ruminants have several components to their digestive system that allow them to digest grass and hay. Bacteria live in a ruminant's stomach and allow the animal to digest large amounts of fiber. One day scientists may be able to genetically alter the bacteria to make them able to digest wood fiber. This would allow cattle and other animals to be fed on woody fiber that is now going to waste. Perhaps monogastric (simple-stomached) animals such as pigs could be developed that could digest roughages like ruminants. This would mean a much cheaper way to produce pork.

Another possibility may be to develop plants like corn that produce their own nitrogen. Corn requires a lot of nitrogen in order to produce large yields. This fertilizer is expensive to supply. Some plants, called legumes, produce nitrogen through the use of bacteria that live in nodules on their roots. If scientists could

BIOTECHNOLOGY: THE FUTURE OF AGRICULTURE

genetically alter the corn plant (along with other grasses) to allow it to host the nitrogen-producing bacteria, grain could be produced much more efficiently (Figure 24-5).

Crops may also be altered to produce more oil or oil of a better type. Remember from Chapter 10 that some crops are produced for the oil from their seed. Scientists think that in the future these as well as other crops may be altered to produce a much higher percentage of oil. This may go a long way in helping to solve the problems caused by running out of petroleum.

Genetic engineering has the potential to be revolutionary in the production of food and fiber. Scientists estimate that early in the twenty-first century, there will be over six billion people in the world. All of these people will have to be fed and clothed. Through the use of genetically altered plants and animals, the efficiency of producing food can be increased and areas that are currently of little agricultural value can be made productive.

Even though there is much hope and optimism over genetic engineering, there is also widespread concern. Some people are worried about the introduction of genetically altered organisms into the environment. They cite examples of plants and animals that were introduced with good intentions and have escaped and caused problems. They feel that by tampering with nature we may unleash organisms that will be hard to control. Many science fiction movies depict scary creatures that have been genetically altered. People note that the most commonly used organism in gene alteration is the *E. coli* bacteria that normally live in the human intestinal tract. Because the bacteria multiply rapidly and the human body is a host for these bacteria, many people feel that we do not know enough about genetic alteration to allow these bacteria into the environment.

The answer to this dilemma is, at least in part, the regulation of genetic engineering. Scientists conduct research on genetic engineering in closely

FIGURE 24-5
If scientists could alter corn to make it a legume, much less would be spent on fertilizer. This scientist is collecting pollen to use in corn genetic research.

FIGURE 24-6
Facilities used in genetic engineering are designed to prevent the escape of organisms.

controlled laboratory situations. The buildings and equipment are designed to prevent the escape of newly engineered microorganisms into the environment (Figure 24–6). Research may take many years before the determination is made that the organisms will be tested outside the laboratory.

The United States Department of Agriculture (USDA) has developed strict guidelines for testing genetically altered organisms that can come into contact with the environment. In addition, the USDA has developed a set of guidelines that all USDA research agencies will use prior to field testing. A heightened public awareness helps to ensure that those doing research remain diligent and do not become careless. It also helps everyone remain conscious of the fact that problems could result from carelessly conducted genetic research.

Careers in Agriscience

You may not have thought about what sort of work you want to do, but sooner or later you must decide. Consider a career in the nation's largest industry—agriculture (Figure 24–7). From stories in the media, you may think that agriculture is dying out. Nothing could be further from the truth. Agriculture is not dying—it is changing. The changes are exciting and

FIGURE 24-7
Perhaps you should consider a career in the nation's largest industry. There is a large need for trained technicians to aid in plant research.

BIOTECHNOLOGY: THE FUTURE OF AGRICULTURE

FIGURE 24-8
Agriculturists work in a variety of settings.

challenging. Remember that all the people in the world will have to be fed and clothed. This will be done through agricultural science.

Consider all the things you like to do. A substantial part of your life will be spent working, and if you enjoy your work, your life will be more pleasant. If you like to work with plants or animals and if you like science, a career in agriculture might be for you. Some people have the mistaken idea that a career in agriculture is difficult, strenuous, dirty work that has no challenge. Today's agriculture can offer exciting challenges. You may work in a variety of places ranging from a food processing plant to a laboratory (Figure 24-8).

SUMMARY

As with any other career, you will need the proper preparation for a career in agriculture. The best place to begin is with the high school Agricultural Education program. In this program you will decide which areas you wish to specialize in. Throughout this book are examples of activities you can get involved in through the FFA. The opportunities are limitless. You can even earn scholarships to help pay for a college education. Agriculture is taught as a two-year program in community and technical colleges, or as part of a four-year university program. These institutions will be glad to send information on opportunities in all their degrees and programs.

24 CHAPTER REVIEW

Student Learning Activities

1. Think of a problem facing agriculture today or in the future. Write a paper giving your ideas as to how the problem could be solved using biotechnology. Be sure to include possible hazards.
2. Interview five people and get their perceptions of genetic engineering. Do you feel their perceptions are correct? Report to the class on your conclusions. Be sure to keep the identity of the people you interviewed confidential.
3. Research the career opportunities in biotechnology. Determine the job responsibilities of a particular career and the qualifications needed. Report to the class on your findings.
4. Search the Internet and find five positive aspects of genetic engineering and five negative aspects. Which aspects are the most credible?

True/False

1. _____ Most changes in agriculture occurred during the twentieth century.
2. _____ Dams and sires are chosen through the use of computers.
3. _____ Scientists have not yet discovered how to clone animals.
4. _____ The majority of insulin is harvested from the pancreases of cattle.
5. _____ Corn requires a lot of nitrogen in order to produce large yields.
6. _____ Scientists estimate that early in the twenty-first century there will be over six billion people in the world.
7. _____ Legumes produce nitrogen through the use of bacteria that live in nodules on leaves.
8. _____ The nation's largest industry is agriculture.
9. _____ Agriculture is not dying; it is changing.
10. _____ A career in agriculture may involve working in a variety of places such as a food processing plant or a laboratory.

✳ Multiple Choice

1. One industry that can never die out is
   a. agriculture
   b. computer production
   c. engineering

2. Perhaps the earliest biotechnology was used to make wine and
   a. bread
   b. apples
   c. cheese

3. One of the newest forms of biotechnology is
   a. insecticide production
   b. manufacturing herbicides
   c. genetic engineering

4. One of the first uses of genetic engineering was in manufacturing
   a. cotton
   b. human insulin
   c. pesticides

5. Genetic scientists have been able to develop plants that are resistant to
   a. herbicides
   b. biotechnology
   c. predators

6. Bacteria live in a ruminant's stomach and allow the animal to digest large amounts of
   a. fiber
   b. biotechnology
   c. grain

7. The best place to begin preparing for an agricultural career is with an Agricultural Education program in
   a. preschool
   b. elementary school
   c. high school

8. The most profound changes in the future will occur in
   a. building construction
   b. genetic engineering
   c. animal care

9. Food, shelter, and clothing are all directly dependent on
    a. agriculture
    b. crops
    c. water

10. Substances that control growth and other bodily functions are
    a. vaccines
    b. hormones
    c. herbicides

Discussion

1. Why is cattle insulin no longer used to help diabetics?
2. Why do scientists want to genetically alter the bacteria found in a ruminant's stomach?
3. What impact does genetic engineering have on plants?
4. Why is there concern about genetic engineering?
5. How is genetic engineering being regulated?
6. How can you prepare for a career in agriculture?
7. Why will agriculture always exist?
8. What are some current uses of genetic engineering?
9. What are some future uses of genetic engineering?
10. How was biotechnology used in the past?

CHAPTER 25
High School Agricultural Education Programs

Student Objectives

When you have finished studying this chapter, you should be able to:

- List the advantages of studying Agricultural Education in high school.
- Name the three parts of the Agricultural Education program.
- Describe the different types of Supervised Agricultural Experience programs.
- Discuss the mission of the FFA.
- Discuss how Agricultural Education is different from the old Vocational Agriculture programs.

Key Terms

Agricultural Education
Vocational Agriculture
Supervised Agricultural Experience
National FFA Organization
Career Development Event
proficiency award

Modern Programs in Agriculture

If you have enjoyed studying the chapters in this book, you will probably want to consider studying more about agriscience. When you enter high school, an entire program centered on the world of agriculture will be available to you. In every state in the country, there are local programs of Agricultural Education designed to help you understand agriculture and to help you develop many different types of skills.

For over 80 years, these programs have helped high school students learn about themselves and the world all around them. The programs began in 1917 and were designed for rural students who lived on a farm. Back then, the program was called Vocational Agriculture. The curriculum and activities were designed to help solve problems faced on the farm. The program also helped students to develop life skills such as social abilities. Over the years, the program has changed a lot. It is no longer a program aimed at students who live on farms (Figure 25–1).

In fact, according to the National FFA Organization, more students enrolled in the program live in urban or suburban areas than live in rural farm areas. Even the name has changed. In 1988, the name was changed from Vocational Agriculture to

FIGURE 25–1

Agricultural Education is not just for students who live in rural areas.

Agricultural Education. The new name described the more modern activities. Agricultural Education is made up of three parts—classroom and laboratory activities, Supervised Agricultural Experience (SAE), and FFA. In the classroom and laboratory, you will learn about a wide variety of topics. The classroom instruction will introduce you to an area of study. Here you may learn about the theory behind a topic. In the lab, you will have hands-on activities to learn the practical applications. Popular areas of study include forestry, horticulture, agriscience, wildlife and natural resources, as well as several other areas. Classroom and lab study in Agricultural Education may be a little different from what you are used to (Figure 25–2). Here you will get an opportunity to actively participate in learning. This means that you will learn by doing. The laboratory may be a little different from what you think of as a lab.

The agriculture lab may be a greenhouse, a well-equipped mechanics laboratory, or a forestry plot. It may be a pond where you learn about raising fish or it may be a nursery where you learn to propagate plants. You will grow plants in the greenhouse, take and analyze water samples, estimate the amount of lumber in a growing tree, grow bacterial cultures, taste dairy products, or wire a light switch. Each day you will have different and interesting activities to do. The science and math concepts learned in other courses will come alive in agriculture class!

Supervised Agricultural Experience

Learning about the exciting world of agriculture does not end when the school day is over. Through the Supervised Agricultural Experience (SAE) component of Agricultural Education, you can be involved in interesting activities after school. One of the best parts is that you can pick the activity that interests you. Your teacher will assist you in determining the

FIGURE 25-2
The Agricultural Education classroom and lab may be different from what you expect.

FIGURE 25-3
Your teacher will work with you after school in a Supervised Agricultural Experience (SAE).

FIGURE 25-4
In an exploratory SAE, you might work with a veterinarian to learn about the duties he or she performs.

correct program for you. Once you have decided, your teacher will help you get started and visit with you periodically (Figure 25-3).

During the visits, he or she will help you to solve any problems and provide any needed assistance.

Types of SAEs

There are several types of SAEs:

Exploratory—This type allows you to explore different areas of interest. For example, you might want to investigate the duties of a veterinarian. You might do research through the Internet, read articles about the profession, or talk to veterinarians about what they do on the job (Figure 25-4).

The next step would be to visit with the veterinarian on the job and maybe even assist in some of the duties. This will give you an idea about whether this would be a good career for you. Exploratory Supervised Experience Programs are very broad based. Your agriculture teacher will be open to almost any idea you may have for further exploring the world of agriculture.

Research/Experimentation—For your SAE, you may want to learn to do scientific research. Your teacher knows how to set up experiments and can teach you how (Figure 25-5). You will learn to use the scientific method.

FIGURE 25-5
In a research SAE, your teacher will help you set up and conduct scientific research.

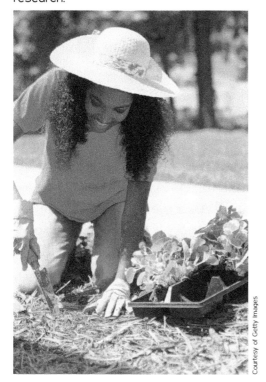

FIGURE 25-6
With an ownership/entrepreneurship SAE you might own a lawn care or landscape business.

You will also learn how to communicate the findings from your research. This type of SAE can be a very small experiment or it can grow into research that you conduct all the time you are in high school.
Ownership/Entrepreneurship—You may wish to start an enterprise of your own. Many successful businesses began as SAEs in high school agriculture programs. You might begin by operating a lawn-mowing business. This may develop into a full-blown lawn maintenance business. It might even evolve into a landscaping enterprise (Figure 25-6).

Many students own livestock. They might begin with a heifer, gilt, or filly, and develop a livestock business. With this SAE, you have the potential for earning money when you are in high school.
Placement—There are many places where you can obtain experience with people in agriculture. You may even get paid for working in a part-time job where you can gain the experience. However, placement SAEs are much more than part-time employment. Although you may earn money, the main objective is to gain experience. You might work in a nursery, on a horse farm, in a tractor dealership, or many other places (Figure 25-7). Your teacher can help you set up the placement SAE and help you design the experience.

The National FFA Organization

Throughout this text there is an insert in almost every chapter that discusses some aspect or program of the FFA. The FFA, which began in 1928, was known as the Future Farmers of America. In 1988, the delegates to the national convention began in 1928 in Kansas City voted to change the name to the National FFA Organization. This new name better represented the broader programs of the Agricultural Education program. The delegates felt that the organization was for students interested in all aspects of agriculture. As a result of the changes in the program and the name, the membership has continued to grow.

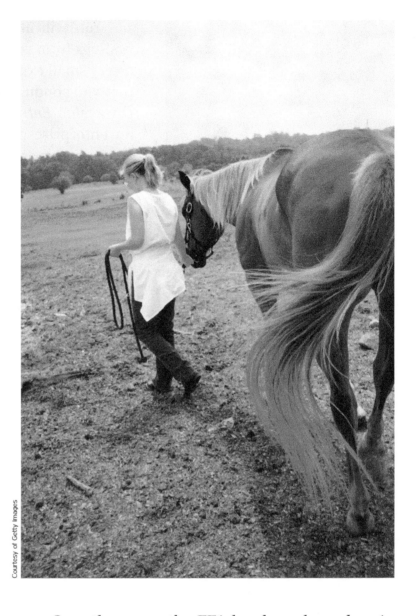

FIGURE 25-7
A placement SAE might involve working on a horse farm to learn about caring for and managing horses.

Over the years, the FFA has been legendary in providing opportunities for young people. The organization is much more than a means to learn about agriculture. In the FFA you can learn a lot about personal development. This means that you will be able to travel, meet many people, learn to communicate, and develop leadership skills (Figure 25–8).

Many famous people credit the FFA with teaching them basic skills. For example, former President Jimmy Carter was a member in Georgia. Speaking about the FFA, he said, "I began to learn how to make a speech. And I began to learn how to work with other people. I also learned the value of agriculture farm families, stability, commitment, idealism,

HIGH SCHOOL AGRICULTURAL EDUCATION PROGRAMS

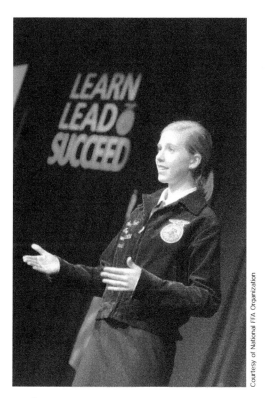

FIGURE 25–8
In FFA you will be able to travel, meet many people, learn to communicate, and develop leadership skills.

hope, truth, hard work, and patriotism from the FFA" (National FFA Organization). Jim Davis, who writes the Garfield cartoon, was an FFA member in Illinois. Senator Sam Brownback from Kansas was a national officer in the FFA. Auburn running back and Heisman Trophy winner Bo Jackson was an FFA member in Alabama.

Today the FFA is a national organization with almost one-half a million members. There are members in all 50 states as well as in Puerto Rico and the U.S. Virgin Islands. The National FFA Organization lists the following characteristics of membership:

In 2009, there were 507,763 FFA members, aged 12–21, in 7,439 chapters in all 50 states, Puerto Rico, and the U.S. Virgin Islands

Ninety-one percent of FFA members are in grades 9–12; 4 percent are in grades 6–8; 5 percent are high school graduates

About 30 percent of FFA members live in rural farm areas; the other 70 percent live in rural nonfarm (39 percent), and urban and suburban areas (34 percent)

FFA chapters are in 10 of the 15 largest cities, including New York, Chicago, and Philadelphia

The top five membership states are California, Texas, Georgia, Oklahoma, and Ohio

Each year more than 50,000 members, advisors, and supporters attend the National Convention in Indianapolis

FFA Programs

The National FFA Organization is dedicated to making a positive difference in the lives of young people by developing their potential for premier leadership, personal growth, and career success through agricultural education. This statement best describes what the FFA is all about. You will make friends and learn about working with others. All these things can be accomplished by participating in the many programs of the FFA.

488 CHAPTER 25

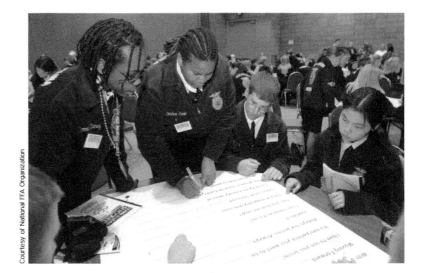

FIGURE 25-9
Chapter activities involve everyone in the chapter and may include competition with other chapters.

CAREER DEVELOPMENT EVENTS

The National FFA Organization offers a wide variety of leadership opportunities for students to get involved in and provides many ways to explore agriculture. The FFA Global Programs offer students and teachers the opportunity to explore international agriculture through programs that teach about values, traditions, lifestyles, and the agriculture of other countries. FFA Global offers members a variety of educational experiences ranging from travel seminars to host-family stays and international internships.

Traveling and learning in the FFA increases awareness of agricultural education in other countries while providing opportunities to develop programs and leadership. As students become familiar with cultural issues, they gain a better understanding and appreciation of others within their country. There are many ways that students and local chapters can get involved in FFA Global programs, such as: chapter-to-chapter exchanges, group travel seminars, the home-stay program through the National FFA, conferences or expositions, hosting a foreign student through an FFA International Inbound Program, or developing an international component to an SAE.

International experiences can extend after high school as well. As students prepare to continue their studies and future careers, they can become immersed

The FFA has several different types of programs. These include chapter activities, team activities, and individual activities (Figure 25-9). Most of the activities involve competitions on the local, state, and national levels.

Chapter activities involve competitions such as the scrapbook competition. The National Chapter Award allows your chapter to compete with other chapters in your state and in the nation. The competition is based on the activities in which your chapter is involved.

Most team activities are part of what is known as Career Development Events (CDEs). These events allow you to use the knowledge and skills you have learned in the classroom and in your

The FFA Global Programs offer students the opportunity to experience lifestyles, traditions, and agriculture in different countries. *Courtesy of Getty Images.*

in global issues through a variety of opportunities. International knowledge and experiences can be gained by internships in international areas around the world [USDA-FAS (Foreign Agricultural Service) Internship Program], International Leadership Seminars for State Officers (ILSSO), World Experience Abroad (WEA), seminars or home-stay programs, and presentations on past international experiences to educate others. More information on international opportunities for students and teachers can be found online at http://www.ffa.org.

FIGURE 25-10

In FFA there are 23 different areas of team competition.

SAE. There is team competition in 23 different areas, including livestock evaluation, dairy products, forestry, and nursery/landscape. Not only can you learn a lot about the skill area, but also you can learn a lot about teamwork and getting along with others (Figure 25-10).

In addition, you will get to travel around the state and to the national convention in order to compete.

Individual activities are those you participate in as an individual person. For example, there are 45 different proficiency award areas. A **proficiency award** is based on your SAE program. Once you decide your area of interest in a Supervised Agricultural Experience program, you can take the program as far as you can. If you excel, you may receive an award for your accomplishments. Proficiency awards carry some nice monetary prizes for the winners.

SUMMARY

No matter what career path you may choose, the Agricultural Education program can help you gain skills that will help you achieve your goals. The three parts of the program (classroom/lab, SAE, and FFA) will provide you with many opportunities to learn and apply knowledge and skills. Experiences you have in the FFA will lead to skills that will be of benefit to you no matter what you do for a career. Don't forget that one of the main advantages of the Agricultural Education program is that it makes learning fun.

25 CHAPTER REVIEW

Student Learning Activities

1. Make a list of the topics in this text that interest you the most. Share your list with the class. Explain why these topics interest you.
2. From the list you made in the first activity, design an SAE. What type of SAE is best for you? What do you expect to learn from the experience?
3. Visit the FFA Web site at http://www.ffa.org. What activities appeal to you?
4. Visit a high school Agricultural Education program. Ask the teacher to explain the activities of the program. Find out which of these activities appeal to you.

True/False

1. _____ Only a few states still have high school agricultural programs.
2. _____ More students enrolled in Agricultural Education live in urban/suburban areas than in rural areas.
3. _____ The name of the program is Vocational Agriculture.
4. _____ The name of the Future Farmers of America has been changed to the National FFA Organization.
5. _____ The three parts of the program are classroom/lab, SAE, and FFA.
6. _____ SAEs include only those activities related directly to farming.
7. _____ Agriculture teachers work with their students after the school day has ended.
8. _____ One of our nation's past presidents is a former FFA member.
9. _____ Two of the top membership states are California and Texas.
10. _____ Agricultural Education has programs of interest to almost everyone.

Multiple Choice

1. The agriculture lab may be
    a. a well-equipped mechanics laboratory
    b. a greenhouse
    c. both a and b

2. The Future Farmers of America was organized in
    a. 1933
    b. 1928
    c. 1917

3. Which former U.S. president was an FFA member?
    a. Herbert Hoover
    b. Bill Clinton
    c. Jimmy Carter

4. FFA began in
    a. St. Louis
    b. Louisville
    c. Kansas City

5. There are FFA members in
    a. all 50 states as well as Puerto Rico and the U.S. Virgin Islands
    b. only states with large rural populations
    c. 43 of the 50 states

6. Exploratory, research, ownership, and placement are types of
    a. FFA programs
    b. classroom activities
    c. SAEs

7. SAE stands for
    a. Standard Agricultural Experience
    b. Supervised Agricultural Experience
    c. Special Agricultural Experience

8. An award based on the activities in which your chapter is involved is called a
    a. Career Development Award
    b. National Chapter Award
    c. Team Activity Award

9. Agricultural Education was once called
    a. Vocational Agriculture
    b. Agriculture Programs
    c. Studies in Agriculture

10. _____ are based on your SAE program.
    a. National Chapter Awards
    b. Proficiency awards
    c. Career development events

Discussion

1. What are the three parts of an Agricultural Education program?
2. Name three types of laboratories used in Agricultural Education.
3. Name the four types of SAEs.
4. Why did the Future Farmers of America change its name to the National FFA Organization?
5. Name three famous people who are former FFA members.
6. How many FFA members were there in 2009?
7. What is a career development event?

CHAPTER 26
Careers in Agricultural Science

Student Objectives

When you have finished studying this chapter, you should be able to:
- Identify the main job classifications in agriculture.
- Analyze the career opportunities in agriculture.
- Identify some of the characteristics associated with various jobs in agriculture.

Key Terms

agribusiness
agronomists
entomologists
horticulturists
geneticists
food scientists
microbiologists

FIGURE 26-1
More than 20 million workers are employed in processing, storing, and distributing food and fiber. This is a large cheese factory. Think of all the people involved in producing cheese and getting it to your table.

During the first half of the twentieth century, agriculture was mostly limited to farming, and preparing for a career in agriculture was fairly easy. People did not think that a formal education was necessary for a farmer. Today, successful people who work in agriculture need a good education in business and in science. Over the years, the number of farmers in the United States has declined dramatically, yet agriculture remains a major source of jobs for Americans. Agriculture professionals are not just farmers and ranchers. Opportunities in this field include areas as diverse as science, engineering, finance, marketing, and exporting. Careers in these areas have expanded as the production and processing of food and fiber have become more specialized and technological. Successful producers rely heavily on the support of these workers. Individuals working in agriculture today are as likely to work in the city as on the farm.

The business of agriculture, or agribusiness, is an important part of the U.S. economy. More than 20 million workers are employed in processing, storing, and distributing the food and fiber produced by farmers (Figure 26-1). Agribusiness also involves manufacturing and selling farm supplies and equipment. The continued success of agriculture depends on the support of this entire network of people who enable the producer to function more efficiently. It is their initiative and hard work that make American agriculture competitive in the world market.

Plant Science

Plant science deals with the growing of fruits, vegetables, and ornamental plants. Career opportunities in plant science vary widely, but all of them are involved in producing plants. The main categories in plant science are horticulture, agronomy, landscaping, and turf management.

The production of field crops requires the expert knowledge of many scientists. Agronomists provide producers with information about the combination

FIGURE 26–2

Crop scientists work to increase the yield of crops by developing new plant strains.

of soils, plants, and climate that will promote the best possible yield. Entomologists provide information about pest control. Today insect populations are not always controlled with pesticides. Parasitic or predator insects can be released in a field to control a pest population. Crop scientists work to increase the yield of crops by improving farming methods and developing new plant strains (Figure 26–2). Some crop scientists study the breeding of crops and use genetic engineering to develop crops resistant to pests and drought. All of these agricultural scientists must have advanced college study in their chosen fields.

Horticulture deals with the cultivation of fruits, vegetables, flowers, trees, and shrubs. Some horticulturists find new ways to grow plants or breed better varieties. They experiment with different varieties of fruits, vegetables, and nuts to make them more nutritious. Others raise and tend plants used for ornamental purposes, such as flowers, shrubs, trees, and grasses. An increase in the interest for indoor and outdoor plantings has raised the demand for these services. People want attractive grounds around their homes and offices, as well as amusement parks, gas stations, fast-food stores, and shopping malls. Skilled horticulturists are needed to grow and tend plants in all of these areas.

Many jobs in plant science require two years of study or less (Figure 26–3). Nursery operators raise and sell flowers, shrubs, hedges, and trees. They are involved in tasks including plant propagation through seeds and cuttings, preparation of soil in outdoor growing areas, greenhouse management, and the storage and packing of plants.

Landscape architects design outdoor areas for people to use, such as parks or gardens, housing projects, or the grounds surrounding hotels. They plan their arrangements to make the best use of the land while protecting the natural environment. They prepare detailed maps and plans showing the placement of trees, shrubs, and walkways. The landscape architect also supervises any grading or construction that is necessary to complete the work.

CAREER DEVELOPMENT EVENTS

The National FFA Convention was founded in 1928, and until 1999 FFA members from all over the country met in Kansas City. In 1999, the convention was moved to Louisville, KY. The convention has grown to become the world's largest student educational convention. Students from every state in the country, plus students from several foreign countries, attend. All types of Career Development Events are conducted during the convention, ranging from public speaking to agriscience. Tours to educational sites are conducted almost every day. Students have a wide array of leadership development workshops to choose from, where they can interact with students from all over the country.

At the convention, students receive awards for many activities that began on the local level and progressed through to the national level. It is quite a good feeling to know that they have earned a national award through hard work and dedication. The FFA awards over a million dollars in scholarships each year.

One of the many highlights of the convention is the gigantic National Agricultural Career Show. Companies from all over the world set up booths to tell about their company. Students can learn about thousands of job opportunities involved with

CAREERS IN AGRICULTURAL SCIENCE

FIGURE 26-3
Many careers in turf production/maintenance and landscaping require less than a four-year college degree.

The National Agricultural Career Show at the National FFA Convention provides information for students on hundreds of different careers in agriculture. *Courtesy of Bill Stagg, National FFA Organization.*

agriculture. Many colleges from all across the United States operate booths where students can get information about various programs and scholarships. They can go home with stacks of printed materials about job opportunities.

Most of the students attending the convention have earned their way. As you begin to study agriculture in high school, set your goal to attend the National FFA Convention. Your teacher can help you design a plan to achieve your goal. Remember, the opportunities in FFA are limited only by your ambition and willingness to set goals and work toward them.

Turfgrass technicians may work in commercial lawn maintenance, private lawn care, or the maintenance of parks, highways, and playing fields. They may be employed as greenskeepers for golf courses, work for the government maintaining public lands and parks, or work in their own businesses. They provide numerous services that include mowing and fertilizing, as well as insect, weed, and disease control.

Animal Science

Careers in animal science are involved with many phases of animal production. These professions include producers, feeders, marketers, manufacturers, distributors, and retailers. They may also include producing pharmaceuticals or equipment that aids in the management of animals. Many of these careers are highly science oriented. The intensive health practices that are necessary to maintain a successful operation in cattle ranching or dairy farming provide many job opportunities in animal welfare. Scientists study the nutrition, growth, and development of domestic animals (Figure 26–4). **Geneticists** look for ways to develop high growth-rate animals that result in a quality product that meets consumer standards. These positions require a four-year program of college study that emphasizes the sciences.

FIGURE 26-4

Animal scientists study the nutrition, growth, and development of domestic animals. This scientist is researching levels of bacteria in the digestive system of cattle.

Cattle reproduction has opened the door to another career in the animal sciences. Artificial insemination is the most common cattle-breeding practice today. Artificial insemination technicians typically have two years of college education. Feedlots also present career opportunities in animal science. At the feedlot the cattle are fattened up so that their market value improves. The manager of a feedlot is responsible for minimizing the costs of feeding the cattle in order to obtain the highest profit. Feedlot operations use computers to regulate feed supplements and medical products. Many feedlot managers are producers who grow their own hay and grain.

Cooperative extension agents provide services to both producers and apartment dwellers. Agents are able to advise people on topics as diverse as crop and livestock production, conservation, home economics, child care, farm management, and nutrition. Cooperative extension agents can answer questions about when to transplant seedling tomatoes or how to remove stains from clothing. Agents often specialize in one of these areas. Extension agents are employed jointly by county and state governments. They must have a four-year degree in one of the fields that relates to extension work.

If you enjoy being around animals you might want to consider pursuing a career in one of the many areas dealing with companion animals. The most important attributes you need for a career in companion animal care are respect and love for animals.

Pet care workers provide a wide variety of services to the owners of small animals. They work in animal hospitals, boarding kennels, and animal shelters. Responsibilities will vary, but may include feeding animals and cleaning and disinfecting cages and equipment. Pet care workers can also work in pet grooming parlors, pet training schools, and pet shops. Many of these positions are entry level and require no special education. Employers often hire high school students and graduates to fill these positions. Some experience with animals is helpful.

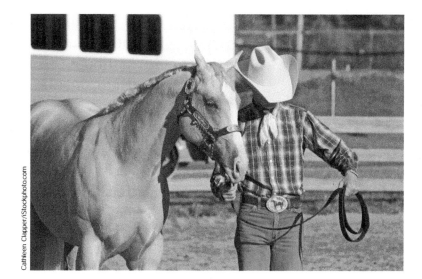

FIGURE 26-5
People who train animals are also in an agriculturally related field.

Animal trainers teach animals to obey commands, compete in shows or races, or perform tricks. Many companion animals can be trained, including dogs, horses, parakeets, and cockatoos (Figure 26-5). Animal trainers can teach dogs to perform a variety of tasks for humans, including the protection of property. Employees in this field may also be involved in training assistance dogs for the blind, the deaf, or the physically challenged. Most animal trainers begin their careers as pet care workers where they gain experience caring for animals.

Dog groomers wash, comb, cut, and trim the fur of dogs and cats. To become a dog groomer it is best to take a course from one of the 50 schools accredited by the National Dog Groomers Association. To enroll in the school, you must be at least 17 years old. Some animals may be nervous, fidgety, or aggressive, so patience and persistence are necessary characteristics of a potential dog groomer.

Horse industry workers breed, train, and care for horses. People who work with racehorses usually begin their careers as grooms. Grooms are responsible for brushing and combing the horses. They are often in charge of feeding the animals and cleaning the stables where they are kept. Grooms may also be involved in training young horses.

FIGURE 26-6
Animal health technologists may assist with laboratory work.

Horse trainers are responsible for determining the best diet and exercise program for the horses. The trainers often supervise grooms and other employees who exercise and care for the horses. Nearly all horse industry workers learn at least some of their skills on the job.

Animal health technologists work side by side with veterinarians. They assist veterinarians with laboratory work, radiology, anesthesiology, and surgery (Figure 26–6). They may also be responsible for animal nursing, sample collection, and general office management. An animal health technologist is not limited to work in a veterinary clinic or laboratory; he or she may also work in a zoo or a circus. These technicians are graduates of a two-year program from a community or technical college.

Veterinarians are animal doctors. They perform a variety of duties in a number of different environments. Veterinarians diagnose animal illnesses, treat diseased and injured animals, perform surgery, and prescribe and administer drugs and vaccines. They also give animal owners advice on the care and breeding of animals. They can work in a clinic, conduct research for the government, or work for a large organization. They can treat both large and small animals, or they can specialize. Almost one-third of all veterinarians specialize in the care of small animals.

Veterinarians must complete four years of college followed by an additional four years of study at a veterinary school. To become licensed to practice, veterinarians must also pass their state's oral and written licensing examination.

Food Science

The food industry is one of the largest industries in the nation. It offers a wide range of opportunities because of the variety of operations and services it performs. Only a small portion of the foods we consume are fresh. Most have been processed by the time they reach the local supermarket. Before a food product can be sold to the consumer, it must be processed, inspected, packaged, labeled, stored, and transported. All of these jobs are part of the extensive processing of food products, from the time the crop is harvested until it reaches the consumer.

One of the main purposes of processing foods is to keep them from spoiling. Processed foods are canned, waxed, dried, frozen, bottled, and pickled to preserve them. Another purpose for processing is to enhance the nutritional value of the food. Highly technical methods of achieving these goals have been developed that require the skills and knowledge of food scientists, microbiologists, and food chemists. All of these positions require advanced college study.

Food scientists improve methods of freezing, canning, storing, packaging, and distributing food (Figure 26–7). They study the chemical changes that take place in food when it is stored or processed. Food scientists try to find ways to process food so that fewer nutrients are lost. They also study the effect of food additives. Many food scientists are attracted to research and development. The most thrilling task of the food scientist is the development of new products. This is a highly competitive area. Developing a product usually takes more than a year and a substantial financial investment by the processing company. Only one product out of ten will survive the

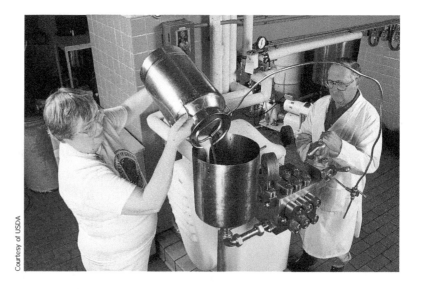

FIGURE 26–7

Food scientists improve methods of storing, packaging, and distributing food.

first year in the supermarket. Ready-to-cook convenience foods continue to account for a major part of the new product development efforts. Thanks to convenience foods, people are able to spend significantly less time in the kitchen preparing meals than they did just a decade ago. Frozen, precooked meals that come in microwaveable packages are a good example of the innovations that have occurred in the food industry in recent years.

Careers in meat processing often deal with the protection of processed meats. Microbiologists lead the way in food safety research. They help the industry find ways to prevent bacterial growth. Food chemists work in the meat industry to ensure a healthful, tasty product. They may study the chemistry of fats in chicken skins to help reduce the cholesterol in human diets, or the muscle chemistry of meats that causes some cuts to be tender and others to be tough. Meat inspection is another career that focuses on a safe product for the consumer, whether it is poultry, red meat, or fish (Figure 26–8). Inspectors who work for the U.S. Department of Agriculture inspect and grade almost all commercially marketed meats.

Food technicians graduate from a two-year program of study. They work in food processing plants and in laboratories. They usually help with quality control, testing, or research under the supervision

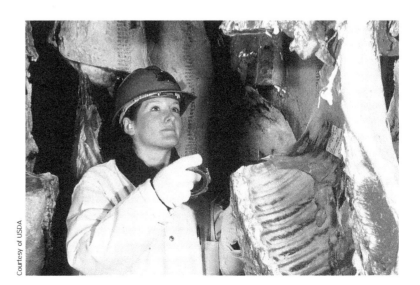

FIGURE 26-8

Meat inspection is a career that focuses on a safe product for consumers.

of a food scientist. Technicians run tests to check processed foods for bacteria, impurities, or poisons. They use instruments that check the quality of the food, including the taste, smell, and color. Some food technicians work as production supervisors and managers in food processing plants.

A number of careers in the food industry deal with agricultural marketing. Agricultural marketing involves the storing, transporting, or marketing of food products. Supermarkets are dependent on the efficient functioning of the agricultural marketing sector. Transportation specialists use all of the forms of transportation available, including trucks, trains, and barges, to get the lowest price possible while maintaining the quality of the product. These specialists must have a good working knowledge of the state and federal laws and regulations that govern the transportation of these products.

Grain merchants are involved in all aspects of buying and selling grain. They act as a connection between the producer and the ultimate consumer. They are concerned with the quality, the shipping, and the processing of the grain. Grain merchants buy most of their grain from area producers at local grain elevators. Merchants must be knowledgeable about the current market prices for grain. Many use computers to get up-to-date prices. Although there are no specific education requirements for grain merchants, some college training is helpful.

The wholesaler is a link in the marketing chain between the producers and the retail store in which the goods will be sold. The wholesaler provides a valuable service by bringing these two segments of the marketplace together. Wholesalers also buy goods in the United States for sale abroad, as well as buying products from foreign markets for sale here.

The retail store manager who buys the food product from the wholesaler is responsible for every phase of the store's operation. Supermarket managers can be responsible for more than 15,000 food and

FIGURE 26-9
Supermarket managers can be responsible for more than 15,000 food and consumer items.

consumer items (Figure 26-9). They also supervise other employees, manage the inventory, account for expenses, and maintain a good relationship with the consumer. The retail manager must constantly evaluate how well each product is doing in order to make effective decisions about when to eliminate one product from the shelf to make room for a new one. This type of evaluation is often made possible by electronic devices that read product bar codes. There are no specified educational requirements for wholesalers and retailers, but a college education is highly recommended.

Natural Resources

In recent years, people in agriculture have been challenged to find ways to grow enough crops to satisfy the needs of consumers without damaging the environment. This problem will become more difficult to solve as the demand for food increases. Many people in agriculture are devoted to the protection and conservation of the environment (Figure 26-10). Highly educated and trained people will be needed to protect the nation's forests, wilderness areas, and recreation areas. Most of the careers available in this area require a four-year college degree.

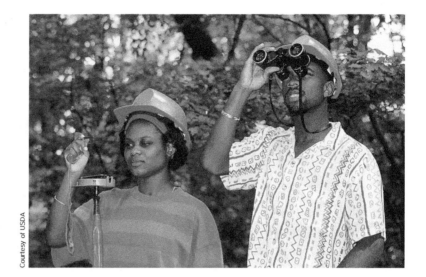

FIGURE 26-10
A career with the forestry service may deal with protecting the environment.

Soil scientists study the characteristics of soil. They gather information and give advice on the management and conservation of soil and water, as well as erosion. Many soil scientists work for the U.S. government. These scientists are called soil conservationists.

Range managers oversee the use of rangelands that have been set aside for livestock grazing and other uses. They supervise the use and conservation of rangelands owned by the government. Range managers balance the grazing needs of ranchers and conservation measures needed to protect the environment. The majority of range managers work for the government. Others may work as appraisers for banks or for companies that have large holdings of rangeland.

Another career that deals with protecting the environment is with the U.S. Forest Service (Figure 26–10). Forest Service researchers conduct studies on wildlife and fish. They study the habitat requirements of individual species to determine how certain activities in the forests and on the rangelands will affect them. The studies allow them to make informed decisions about cutting timber or allowing grazing of the rangelands.

CAREERS IN AGRICULTURAL SCIENCE

SUMMARY

The future of careers in agriculture looks bright. If current predictions hold true, workers will need to become even more efficient because the demand for food and other agricultural goods will continue to increase. If the world's growing population is to be fed, world food production must double in the next 40 years. In addition, changing consumer tastes will continue to prompt new product development. Agricultural research will become more important in the years to come as new technologies are needed to meet the needs of the American consumer while maintaining a healthy environment (Figure 26–11).

FIGURE 26–11
Agricultural research will become more important in the years to come.

26 CHAPTER REVIEW

Student Learning Activities

1. Choose a career that interests you. List as many of the tasks performed in this career as you can. Find out what qualifications are required to be hired in this position.
2. Interview a person who works in the career you have chosen. List all the things he or she enjoys about working in that job.
3. Look in the classified ads in your local newspaper. How many jobs are available that deal with agriculture? What are the qualifications for the jobs?

True/False

1. _____ Agriculture is the same thing as farming.
2. _____ Horticulturists may raise plants for ornamental purposes.
3. _____ Cooperative extension agents only serve producers.
4. _____ Most of the foods that we consume today are fresh.

5. _____ Careers in meat processing often deal with the protection of processed meats.

6. _____ One of the main purposes of processing foods is to alter the flavor.

7. _____ There will be fewer jobs in agriculture in the future.

8. _____ An individual looking for a job in agriculture today will probably work in the country.

9. _____ More than 20 million people are employed in agribusiness.

10. _____ Almost one-third of veterinarians specialize in the care of small animals.

Multiple Choice

1. The business of agriculture is called
    a. horticulture
    b. agronomy
    c. agribusiness

2. Entomologists provide producers with information about
    a. pest control
    b. the best combination of soils, plants, and climate
    c. breeding ornamental plants

3. Individuals who deal with the chemistry of fats in chicken skins are
    a. food chemists
    b. microbiologists
    c. entomologists

4. Horticulturists are experts in
    a. designing outdoor areas for people to use
    b. how to grow, pick, and ship plants
    c. artificial insemination

5. Individuals who give advice on the conservation of soil and water are
    a. soil scientists
    b. agronomists
    c. Forest Service researchers

6. One of the purposes of processing food is
    a. to decrease the nutritional value
    b. to study food additives
    c. to keep it from spoiling

7. A major part of new product development efforts is related to
   a. increased job opportunities
   b. convenience foods
   c. the study of food additives

8. Individuals who act as links in the marketing chain between the producers of goods and the retailer are
   a. food technicians
   b. retail store managers
   c. wholesalers

9. A landscape architect would be involved in
   a. designing the grounds surrounding an apartment complex
   b. mowing and fertilizing the grounds of a golf course
   c. the cultivation of plants

10. Job opportunities in agribusiness have increased because of
    a. the increase in the number of farmers in agriculture
    b. an increase in the amount of land that is planted
    c. an increase in the specialization of the processing of food and fiber

Discussion

1. How has the role of education changed for farmers in the last 50 years?
2. How does agriculture continue to be an important part of the economy if the number of farmers has declined?
3. List three jobs in the area of plant science.
4. Why do producers today depend on a network of people?
5. Name three topics about which a cooperative extension agent would be able to answer questions.
6. What are some of the tasks of a food scientist?
7. What is agricultural marketing?
8. What challenge does the environment present to producers?
9. Why does it look like careers in the field of agriculture will continue to be important in the future?
10. Why is food processed?

Glossary

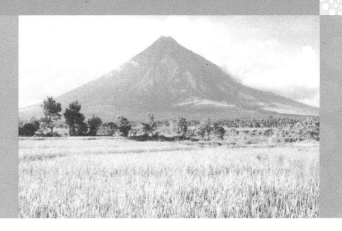

A

A horizon the leached upper portion of a soil profile; the eluvial layer.

Adventitious root a root growing from a stem or branch and not from root tissue.

Aerator a mechanical device used to put oxygen into a pond or lake.

Aerial root a root growing from the stem which is exposed to the air.

Agribusiness the business part of agriculture that deals with economics, marketing, etc.

Agricultural Education a high school and/or middle school educational program that consists of three parts—classroom and laboratory activities, supervised agricultural experience (SAE), and FFA.

Agriculture the industry engaged in the production of plants and animals for food and fiber, the provision of agricultural supplies and services, and the processing, marketing, and distribution of agricultural products.

Agronomists scientists who study the production of field corps.

Alluvial soil soil that is transported and deposited by running water.

Amino acids the building blocks of protein.

Anaerobic bacteria bacteria that thrive in the absence of oxygen.

Animal rights the philosophy that supports the idea that animals have the same rights as humans.

Animal welfare the philosophy that supports the idea that animals should be well cared for and well treated.

Anther the part of the stamen on seed-producing plants which develops and contains the pollen.

Antibody a substance manufactured by an animal's body to combat a disease.

Aquaculture the growing of aquatic animals for human use.

Aquatic animal an animal that lives in water.

Aquifer a geologic formation in rock or sand that contains water.

Artificial insemination (AI) the placing of sperm in the reproductive tract of the female by means other than the natural breeding process.

Asexual reproduction the propagating of plants from a single plant.

B

B horizon the mineral horizon below an A or E horizon.

Balance a basic landscaping principle that no one part of a landscape scheme should be more important than any other part.

513

Bedding plant flower and vegetable plants that are planted in beds.

Beef the meat obtained from cattle. Usually refers to meat from an animal over the age of one year.

Biodiesel an alternative fuel source made by mixing oils from plant and/or animal sources with petroleum-based diesel fuel.

Biofuel a liquid, solid, or gaseous energy source created from organic materials.

Biological control the control of pests using natural means such as insect predators or weed eating insects.

Biologist a person who studies living organisms as a career.

Biology the field of study dealing with living organisms.

Biomass a term for large amounts of organic waste products such as lawn clippings, wood chips, or manure. It can be used as a renewable energy source.

Biotechnology the application of biological and engineering techniques to living organisms.

Blanching the process of stopping enzyme action in food by dipping the food first in hot water then in cold water.

Border plants plants established along the borders of a field to attract insects away from the crop.

Boring Tool a device used for boring a hole in wood, metal, plastic, or other material.

Bottom Dead Center (BDC) the position of a piston when it is at the bottom of its stroke.

Bovine somatotropin (BST) a naturally occurring hormone that aids in stimulating milk production in cows.

Brace a tool used for turning a bit that bores a hole in wood. It is turned by hand.

Brackish water that is neither salt nor fresh but contains a low level of salt.

Breed animals having a common origin and distinguishing characteristics.

Broiler a chicken raised for market that is 2.5 pounds or more. They are usually 6 to 8 weeks old.

Brood pond a pond used to keep fish that breed and lay eggs.

Brooder a device used to keep chicks warm.

Bulb the below-ground bud of a plant.

Bull a male bovine that has not been castrated.

Butter a product made from milk fat.

By-product something other than the main or intended product of a process.

C

C horizon the mineral horizon or layer, excluding indurated bedrock, that is little affected by soil-forming processes and does not have the properties typical of the A or B horizons.

Caloric content the amount of energy a food contains.

Cannibalism the eating or killing of animals by their own kind.

Canning a food preserving method that uses sealed cans or jars of food that are exposed to high heat to kill microbes.

Canopy the uppermost growth of crowns of trees.

Carbon cycle plants use carbon dioxide and sunlight in photosynthesis. After the plants die, when the dead plant life is used as fuel, it releases carbon dioxide back into the atmosphere.

Carcass the part of a meat animal remaining after slaughter and the hide, head, feet, and internal organs have been removed.

Carcinogen any cancer-causing agent.

Career Development Event competitive events conducted as part of FFA. The events help develop career skills.

Catalyst a substance that accelerates a chemical reaction, but is not destroyed by the reaction.

Cereal pertaining to grains or products made from grains.

Cheese food made from the consolidated curds that have been separated from milk.

Chisel a long, flat tool that has a very sharp edge used for cutting.

Cholesterol a fat-soluble substance found in the fat, liver, nervous system, and other areas of an animal's body. It plays an important role in the synthesis of bile, sex hormones, and vitamin D.

Citrus fruits fruits from trees of the genus Citrus such as lemons, oranges and grapefruit.

Civilizations a grouping of people who live in an ordered society.

Clamp a term used to describe several types of devices used to temporarily hold material together.

Clay a sediment of soft, plastic consistency composed primarily of fine-grained minerals.

Climax vegetation the plants that are the last to appear as a result of natural succession.

Clone a plant or animal that originates from only one parent.

Closed-loop system a system of tanks where fish are raised. Oxygen is provided and the water is filtered.

Cocoon the covering that surrounds an insect pupa.

Cold frame an enclosed unheated frame used for growing plants.

Combine a self-propelled or tractor-drawn machine that cuts, threshes, and cleans the standing crop while moving across the field.

Combustion the chemical process of oxidation that produces heat.

Commercial fertilizer plant nutrients derived from processing involving chemical mixtures and sold to producers.

Commodity a transportable resource product with commercial value; all resource products that are articles of commerce.

Companion animals animals that are raised to be pets or companions to humans.

Composting the process of piling organic matter together so it will undergo chemical changes that make the matter useable for fertilizer.

Compression ratio a comparison of the volume of an engine cylinder at top dead center (TDC) to the volume of the cylinder at bottom dead center (BDC).

Compression stroke the upward movement of a piston in an internal combustion engine that compresses gases inside the cylinder.

Concentrate a feed high in carbohydrates and low in fiber.

Confinement operation operating systems where animals are confined in buildings or cages.

Conifer an evergreen tree that produces cones.

Consumer a person who buys goods or services from other people.

Controlled environment an environment for animals that is kept at the correct temperature and other conditions to maximize animal comfort.

Cool water fish fish such as trout and salmon that thrive in water below 65 degrees F.

Cotton gin a machine used to separate the cotton seed from the lint.

Cover crop crops established for the purpose of adding organic matter to the soil. They are usually planted in the winter and plowed under in the spring.

Cow a female bovine that has had a calf.

Crop residue stalks and other debris left in the field from last year's crop.

Crop rotation planting a different crop than was planted the last year.

Crossbreed an animal with parents of different breeds.

Cross-pollination pollination that is the result of gametes from different plants.

Crustacean aquatic animals with a rigid outer covering, jointed appendages, and gills.

Cultivars plants that have been developed and culti-vated by humans for their use.

Cultivated plant any plant that has been planted, tended, harvested, and improved.

Curd a milk product consisting of casein and fat; the coagulated part of milk.

Customary Measuring System a measuring system used in everyday life in the United States. Inches, feet, and yards are common measurements in the customary system.

Cut flower a flower that has been harvested by cutting it off with the stem attached.

Cutting A part of a plant that has been cut off for the purpose of propagating a new plant.

Cycle the complete sequence of events that occurs within a combustion chamber of an internal combustion engine.

D

Dairy pertaining to products that are derived from milk. Also, a farm where cows are kept for milk production.

Dairy product any food product that is made from milk.

Dam breed a breed of animal that makes good mothers that are used in crossbreeding programs.

Dam the mother of an animal.

Debeaking the removal of the beak from a baby chicken.

Deciduous refers to a plant that loses its leaves every year.

Diet the amount of food taken in during a 24-hour period. The term can be used in referring to either human food or animal feed.

Disease any departure from the normal health of a plant or animal.

Disinfectant a substance that kills pathogens (germs).

Dissolved oxygen oxygen that is dissolved in water and is usable by animals that breath through gills.

Distillation Purification of a liquid by heating it until it converts to a vapor, then cooling until it converts to a liquid again.

Division A means of plant propagation that involves cutting crowns, rhizomes, tubers, or roots into sections.

Docile Refers to an animal that is gentle in nature.

Docking the removal of an animal's tail. This is done to help keep the animal clean or to prevent tail biting.

Dressed refers to an animal that has been slaughtered and the carcass cleaned.

Drug residue any amount of a drug that is left in an animal's body.

Drying a process of food preservation where moisture is removed from the food.

E

E horizon a mineral horizon, mainly a residual concentration of sand and silt high in content of resistant minerals as a result of the loss of silicate clay, iron, aluminum, or a combination of these.

E. coli bacteria a bacteria that lives in the colon of mammals. It can cause a variety of diseases.

Ecology the totality or pattern of the interrelationship of organisms and their environment, and the science that is concerned with that interrelationship. The subject is often categorized into plant, animal, marine, terrestrial, freshwater, desert, tundra, and tropical ecology.

Ecosystem the entire system of life and its environmental and geographical factors that influence all life, including the plants, animals, and the environmental factors.

Ectothermic refers to an animal that gets its body temperature from its environment.

Electronic Measurers devices that measure distances electronically.

Embryo transfer the process of removing embryos from superior females and transferring them to inferior females.

Emphasis see Focal point.

Endothermic refers to an animal that generates its own body temperature internally.

Engine a device or machine that can produce power on its own independent from an external source of energy.

Engine timing the correct sequence of ignition, piston movement, and valve opening and closing that occurs in a properly running engine.

Entomologists scientists who study insects.

Environment the sum total of all the external conditions that may act on an organism.

Environmental Protection Agency (EPA) an agency of the federal government that oversees environmental concerns.

Enzyme a substance that speeds up or stimulates a chemical reaction.

Erosion the wearing away of the soil by water or wind action.

Ethanol a form of alcohol that is considered to be one of the most promising alternative fuel sources.

Exhaust stroke the movement of the piston within a cylinder that expels exhaust materials.

Exotic plant a plant that is out of the ordinary.

Experiment station any of the USDA research facilities associated with state agricultural universities, where new ways of farming, ranching, or rural living are officially tested.

Export shipment of commodities to foreign countries.

E10 a fuel made from 10 percent ethanol and 90 percent gasoline.

E85 a fuel made from 85 percent ethanol and 15 percent gasoline.

F

Farm factories a derogatory term used to describe farms that produce animals in an inhumane way as if they were produced in a factory.

Farrow the process of giving birth to pigs.

Farrowing crate a cage or crate in which a sow is placed for farrowing. It keeps her from crushing the pigs.

Feedlot a lot where cattle are kept for fattening for market.

Fermentation the processing of food by the use of yeasts, molds, or bacteria.

Fertilizer any organic or inorganic material added to soil or water to provide plant nutrients and to increase the growth, yield, quantity, or nutritive value of the plants grown therein.

Fiber crops crops such as cotton and flax that are grown primarily for fiber.

Fibrous root a root system which is composed of many fine, branched roots.

Fingerling a fish from 3 to 6 inches long that is used for stocking ponds or lakes.

Fish slurry a material made from the entrails, heads, bones, etc. of fish. It is a rich source of plant nutrients.

Flock a group of poultry or sheep.

Flood plain the area along a river or body of water that is covered in water by periodic flooding.

Floriculture the cultivation of plants for their flowers.

Florist a person who deals in arranging and selling flowers.

Flour the fine-ground product obtained in the commercial milling of wheat and other grains which consists essentially of the starch and gluten of the endosperm. It is used for baking bread, pastry, cakes, etc.

Focal point a particular point in a landscape that draws the observers eye.

Foliage plant a plant that is grown for its attractive foliage.

Food and Drug Administration (FDA) an agency of the federal government that regulates food and drugs that are sold to consumers.

Food chain the transfer of food energy from the initial source in plants through a series of organisms by repeated eating and being eaten.

Food scientists scientist who study the processing, preservation, etc. of food.

Forestry the science of growing trees.

Fossil fuels fuels that originated from plants and animals that existed millions of year ago.

Four-stroke cycle engine an internal combustion engine that requires four complete movements of the piston to complete the cycle.

Free range allowing animals to live in an area unrestricted by a fence.

Freeze-frying a food preserving method that uses a vacuum to remove moisture from frozen foods.

Fresh water water that contains no or very little salt content.

Fruit the matured ovary of a flower and its contents including any external part which is an integral portion of it.

Fry small, newly hatched fish.

Fumigate the destruction of pathogens by the use of liquids or solids that release a vapor.

Fungi plantlike organisms that have no chlorophyll; they get their nourishment from living or decaying organic matter. Plural of fungus.

G

Gaited the definite rhythmic movement of a horse such as trot, canter, pace, etc.

Game bird a bird that is hunted for food.

Gene the simplest unit of inheritance. It is composed of DNA.

Geneticists scientist who study the transfer of genes in the reproductive process.

Genetic engineering the altering of the genetic components of an organism by human intervention.

Geology the study of the Earth, earthy materials, their history, processes, and products. This includes water, air, minerals, and rocks.

Germination the emerging of a plant from a seed.

Glacier a slow-moving mass of ice.

Grafting a process where a new plant is propagated by attaching the root system of one plant (such as an apple tree) with a stem from another plant.

Greenhouse any of several different types of heated, glass- or plastic-covered structures used for the growing of plants.

Groundwater water within the ground that supplies wells and springs.

Growing season the number of days from the average date of the last freeze in the spring to the average date of the first freeze in the fall; the period in which most plants grow.

H

Hammer a tool used for driving nails.

Hand Tool tools that are used and powered by humans.

Hardwood wood from a deciduous tree.

Heavy soil Generally, a clayey soil in contrast with a sandy one. The reference is to resistance and need for greater power in pulling a plow rather than to actual weight or specific gravity which is less for dry clay than for dry sand.

Heifer a female bovine that has not had a calf.

Herbicide a substance that is used to kill plants.

Heterosis the degree of superiority of a crossbred animal over what would be expected of its purebred parents. Also known as hybrid vigor.

Hippotherapy a therapy that uses animals such as horses to help physically challenged people.

Hive a box used to hive a swarm of bees.

Homogenization the process where the fat component of milk is blended with the other components of milk.

Homogenized milk milk that has been processed to blend the butter fat.

Hormone a chemical substance produced by an organism that causes a specific effect on the animal or a portion of its systems.

Horticulture the branch of agriculture that deals with the growing of fruits, flowers, vegetables, or ornamental plants.

House plant an ornamental plant that is grown for its beauty and kept in the house.

Humus organic matter in the soil.

Hunters and gatherers people who obtain their food by gathering wild plants and animals.

Hutch a cage and shelter to house rabbits.

Hybrid vigor *see* heterosis.

Hydrological cycle the cycle of water evaporating and condensing and beginning the cycle over again.

519

I

Import goods that are brought into this country from another country.

Incubator a device that controls the temperature and relative humidity for the hatching of eggs.

Infection the invasion of an organism's tissues by a disease-causing agent.

Inorganic soil a mineral soil; a soil in which the solid matter is dominantly rock minerals in contrast to organic soils, such as peats and mucks.

Insecticide a substance used to kill insects.

Insects organisms that have three distinctive body parts and have wings in at least one stage of their lives.

Intake stroke the movement of the piston within a cylinder that draws fuel into the cylinder.

Integrated pest management a method of controlling pests that uses a variety of means.

Interiorscaping the use of plants inside a building to produce a pleasing effect.

Internal Combustion Engine a machine or device that is capable of converting heat energy into mechanical power. The heat is produced inside of the machine.

Irrigation the artificial application of water to soil for the purpose of increasing plant production.

J

Jerky a meat product made by drying.

K

Kerf the gap cut into a board or other material by a saw.

Koi fish of the carp family that are raised because of their colorful appearance.

L

Lagoon a pond used for disposing of animal manure.

Land grant Act an act of congress passed in 1862 that provide public land to fund the creation of colleges.

Land-grant university any of the state colleges and universities started from federal government grants of land to each state to encourage further practical education in agriculture, homemaking, and the mechanical arts. The mission of these universities is to conduct programs in teaching, service, and research.

Landscape architect a person trained in the art and science of arranging land and objects on it for human enjoyment and use.

Landscape contractor a person who installs a land-scaping plan.

Landscape designer a person who designs and draws up plans for a landscape.

Landscaping to beautify the terrain with the planting of trees, shrubs, and plants.

Landscaping cloth a cloth used to prevent weeds from growing among landscape plants.

Larvae the immature insect as it hatches from the egg.

Layer a hen that is raised for producing eggs.

Legume a family of plants that with the aid of symbiotic bacteria converts nitrogen from the air to build up nitrogen in the soil.

Level a tool that uses a bubble inside an enclosed glass capsule of mineral spirits to indicate levelness.

Life cycle the changes in life form that an organism undergoes.

Litter a group of young pigs that were born from the same mother. Also the absorbent material spread on the floor of an animal's pen or house.

M

Mad Cow Disease a disease in cattle that is caused by a type of protein that attacks the nervous system. Mad cow Disease can be transmitted to humans if they eat poorly cooked beef from an animal with the disease.

Magneto ignition system an engine electrical system in which a spark is generated from a magnetic field created by magnets.

Manure animal waste. It can be used as fertilizer or fuel.

Marking Tool a tool to mark a line on a board or other material.

Mash A cornmeal-and-water mixture used in making ethanol.

Mastitis a disease of the mammary glands that is usually caused by an injury.

Measuring Device any tool used to measure.

Measuring Tape a device that measures linear distances. It is flexible and rolls up into a case.

Media soil or soil-like material in which plants grow.

Meteorology the study of the physical processes which occur in the atmosphere and of the related processes of the lithosphere and hydrosphere.

Methane The largest component of natural gas.

Metric Measuring System a measuring system based on a decimal system, with the meter being the base for measurement. Measuring devices in the metric system are marked as meters, decimeters, centimeters and millimeters.

Microbiologists scientist who study microbes such as bacteria, viruses, etc.

Microorganism an organism so small that it cannot be seen clearly without the use of a microscope; a microscopic or submicroscopic organism.

Milk fat the fat content of milk. Whole milk has about 3.5% fat content.

Mineral a chemical compound or element of inorganic origin.

Motor a device or machine that produces power from the energy supplied by an external source, such as electricity.

Mule an animal produced by the mating of a jackass and a mare.

Mutton the meat of a mature sheep.

N

National FFA Organization a program conducted as a part of Agricultural Education that centers on leadership development and personal growth.

Natural gas a form of gas created by the transforma-tion of plant and animal life into fossil fuel.

Natural resource any resource that grows or occurs naturally.

Natural toxins poisons released by plants as a natural defense against insects.

Navigable waterway a stream or body of water that is large enough to allow the passage of boats or ships.

Nectar the sweet substance from flowers. Bees make honey from this material.

Nematode microscopic, wormlike, transparent organism that can attack plant roots or stems to cause stunted or unhealthy growth.

Node the place on a stem which normally bears a leaf or whorl of leaves.

Nodule a root formation on certain leguminous plants produced by invasion of symbiotic, nitrogen-fixing bacteria. The bacteria furnish the plant with fixed nitrogen compounds and receive nutrient plant juices like carbohydrates.

Nonpoint source pollution that occurs from several sources rather than a single source.

Nonrenewable energy an energy source that is gone forever when its supply is depleted.

Nonselective herbicide a substance that kills all plants it contacts.

Nutrient an element or compound in a soil which is essential for the growth of a plant.

Nutrition any substance such as protein, carbohy-drates, vitamins, etc that is used by an animal to sustain growth, health, reproduction, etc. also, Elements such as nitrogen, phosphorus, potassium, ect. that are used by plants.

O

O horizon organic horizons above mineral soil.

Oil crops crops such as soybeans and canola that are grown primarily for the oil in the seeds.

Orchard a field used for growing fruit trees.

Organic plants or animals that are grown without commercial fertilizers, hormones, pesticides, etc.

Organic fertilizer plant nutrients derived from organic sources such as compost, manure, etc.

Organic matter matter found in or produced by living animals and plants which contains carbon, hydrogen, oxygen, and often nitrogen and sulfur.

Organic soil a soil having a high content of organic soil materials.

Ornamental fish fish that are grown for their appear-ance such as color, body shape, etc.

Ornamental plant any plant produced for its beauty.

Ovary the female organ that produces eggs and female hormones.

P

Parasite an organism that lives off of or at the expense of another organism.

Pasteurization the process of killing microbes in milk by the use of heat.

Pasture an area of grass maintained for animal grazing.

Per capita consumption the average amount of a commodity a single person within a particular population consumes in a given period.

Pesticide any substance used to kill pests.

Pesticide residues tiny amounts of pesticides that are left on produce after harvesting.

Pests organisms that grow where they are not wanted or that do damage to wanted plants or animals.

pH a numerical measure of acidity or hydrogen ion activity of a substance such as food or soil. The neutral point is pH 7.0. All pH values below 7.0 are acid and all above 7.0 are alkaline.

Pharmaceutical any substance used to enhance the health of humans or animals.

Phloem the tissue of a plant that transports food.

Photoperiodic plants that react to specific periods of light.

Photosynthesis the process in plants that converts light energy to chemical energy.

Pickling a method of preserving food that uses an acid such as vinegar as a preservant.

Pistil the female element of a flower.

Plane a hand tool used to smooth boards.

Plant nurseries operations that propagate and grow plants for sale.

Plumb Bob a tool consisting of a weight on the end of a string that indicates whether a material is plum (vertically level).

Point source pollution that occurs from a single source.

Pollination the transfer of pollen from the anther to the stigma of a flower.

Pollution substances in a body of water, air, soil, etc., that impair the usefulness of it.

Pome fruits fruits from trees in the family Rosaceae that have fleshy insides with small seeds. Apples and pears are pome fruits.

Pork the meat from pigs.

Potted plant a plant that lives its entire life in a pot.

Poultry domestic fowls that are raised for meat, eggs, or feathers.

Power a source of energy or a means of completing work.

Power Stroke the movement of the piston within a cylinder that creats power.

Power Tool a tool that has an outside power source such as electricity.

Predator an animal that eats other animals.

Predator insects insects that eat other insects.

Proficiency Award an award given for excellence in Supervised Agricultural Experience Programs (SAEP).

Propagation increasing the number of plants by causing them to reproduce.

Proportion and scale a landscaping principle that components of a landscape should be consistent in size. For example, a large house should have large trees and a small house should have small trees, etc.

Protozoa a group of one-celled organisms that generally do not contain chlorophyll; sometimes classified as one-celled animals.

Pruning the removing of limbs or other plant parts to shape the plant.

Pupa the stage of an insect's life between the larva and the adult stage.

Purebred an animal belonging to one of the recognized breeds. The animal's ancestry are all of that breed.

R

Ration all the feed an animal receives during a 24-hour period.

Refrigeration artificial cooling, either by the application of ice or by utilizing the principle of the latent heat of evaporation.

Registration the process of certifying that an animal is a purebred of a particular breed.

Renewable resource a resource that can be replaced such as trees that are grown as a crop.

Rennet a substance derived from the stomachs of calves and used in the cheese process.

Resistant designating a plant or animal that can resist disease.
Rhizobia bacteria that live symbiotically in roots of legumes and fix nitrogen from the air.
Rhizosphere the root zone, the area where microorganisms are most active in increasing the availability of nutrients for plants.
Rhythm a landscape principle that components of a landscape should be consistent.
Ripping the act of cutting a board lengthwise (with the grain).
Roe a mass of fish eggs.
Roots plant structures that are usually in the soil that absorb water and nutrents for the plant.
Root cutting a cutting taken from a plant root for the purpose of growing a new plant.
Rooting hormone a substance applied to a plant or root to stimulate growth.
Roughage a feed that is high in fiber content and low in carbohydrate content.
Ruler a short tool used to measure lines.

S

Salting a food preserving method that uses salt as a preservant.
Sand any of the small, loose, granular fragments which are the remains of disintegrated rocks. It may contain a great variety of minerals and rocks, but the most common mineral is quartz.
Scavenging the process of moving the exhaust gases from the cylinder of an engine.
Screwdriver a tool used for turning a screw in wood, metal, plastic or other material.
Sea island cotton a woody plant grown as an annual for its long staple lint cotton used in fine cotton fabrics. Native to tropical America.
Seed the embryo of a plant.
Seedling a plant that has newly emerged from the seed.
Selective breeding choosing the most desirable plants or animals for breeding in order to develop superior plants or animals.
Selective herbicide a substance that will kill some plants and have little or no effect on other plants.
Self-pollinating a plant that produces and uses its own pollen for reproduction.
Semen the fluid from a male animal that contains the sperm.
Service animals animals that are raised to provide help for physically challenged people. For example, guide dogs that help sight impaired people.
Sexual reproduction reproduction through the means of the uniting of gametes.
Shade cloth a cloth used to cover plants to reduce the amount of sunlight.
Shrub a plant that has several woody stems and is low growing.
Silage a feed made from chopped and fermented corn.
Silt small, mineral soil particles ranging in diameter from 0.05 to 0.002 mm in the USDA system or 0.02 to 0.002 mm in the International system.
Sire the father of an animal.
Sire breed breeds of animals that are used primarily as sires in a cross-breeding system.
Skim milk milk that has had almost all of the milkfat removed.
Smith-Hughes Act a United States federal law passed by Congress in 1917 which gives federal aid

for vocational education in the secondary schools.

Softwood wood from a conifer tree.

Soil the mineral and organic surface of the earth capable of supporting upland plants.

Soil amendment any material added to the soil that improves it.

Soil horizon a profile of the soil that shows the distinct parts of the soil such as the top soil, sub soil, parent material, etc.

Soil texture the coarseness or fineness of soil.

Spawn the process of a fish laying eggs.

Squab the meat from a pigeon.

Square a tool used for laying out lines and angles.

Stamen the male part of a flower that bears the pollen.

Stems plant structures that provide support for the plant.

Stigma the female part of the flower that receives the pollen.

Stock plants plants grown for the purpose of producing new plants from propagation methods such as cuttings.

Stocker a beef animal that weighs around 700 to 800 pounds that needs additional growth before entering the feedlot.

Stone fruits fruits such as peaches and cherries that have a large, hard seed in the center.

Stillage by-product of ethanol production. It is used in the production of a high-quality livestock feed.

Strip cup a cup used to catch the first milk milked from a cow.

Stroke the up or down distance traveled by a piston in an internal combustion engine.

Succession the replacement of one type of plants for another in the natural progression of a forested area.

Supervised Agricultural Experience an after school educational program connected with a high school Agricultural Education Program that is aimed at gaining practical experience in some area of agriculture.

Surface water water that lies on the surface of the ground.

Symbiotic relationship a relationship between two different types of organisms that is beneficial to both of them.

T

Taproot the main descending root of a plant from which branch roots grow.

Teat the part of the female mammary system from which the young receives milk.

Terminal tip the end of a branch, stem, or root where growth takes place.

Texture the size of the particles in a particular soil sample.

Tissue culture the process of growing a new plant from plant cells.

Tolerant the ability of a plant or animal to resist a pest or disease.

Top Dead Center (TDC) the position of a piston when it is at the top of its stroke.

Topiary the shaping of plants into shapes that are not natural such as into the shape of animals.

Trade balance the balance of imports and exports of a country.

Tree farm a farm that grows trees for commercial use such as pulp, lumber, Christmas trees, etc.

Two-stroke cycle engine an internal combustion engine that requires two complete movements of the piston to complete the cycle.

U

Udder the exterior portion of a females mammal's mammary system.

Upland cotton a tropical, woody herb grown as an annual for its fiber, the source of much of the commercial cotton in the United States.

USDA the United States Department of Agriculture, a department in the Federal Government that oversees matters concerning agriculture.

V

Variety and unity a landscaping principle that all components should be unified by a common component such as a walk way and at the same time have some variation in the plan.

Veal meat from a calf that is less than 3 months old.

Vegetable the edible part of an herbaceous plant.

Vocational Agriculture a program designed for rural youths in which they studied agriculture. The program has been revised and is now called Agricultural Education.

Volcanic ash the fine particles from volcanic eruptions which, in thick deposits, are the parent material of soil.

W

Warm water fish fish such as catfish that thrive in water around 85 degrees F.

Wean to take a young animal away from its mother when it is mature enough to eat on its own.

Weed any plant that is growing where it is not wanted.

Wetlands an area that stays wet for most of the year.

Whey the watery portion of milk that remains after the curd and cream have been removed.

Whorled three or more leaves, twigs, etc., that are arranged in a circle at one point on a plant.

Wilderness area an area set aside by the government to preserve the wild state of the area. Access is limited.

Wildlife habitat the environment in which wildlife lives.

Windbreaks trees that are planted to block the wind from homes, roads, crops, etc.

X

Xylem plant tissue that transports water.

Y

Yogurt a semisolid fermented milk food product.

Z

Zoonose a disease that can be contracted by humans and animals.

Index

A

Adventitious roots, 76
Aerial roots, 76
Agribusiness, 496
Agricultural Education, 482–483
Agricultural marketing, 506
Agriculture
  changes in, 37–38
  described, 4
  environmental protections, 231–232
  importance of, 4–6
  nitrate pollution, 242–244
  organic. *See* Organic agriculture
  soil erosion, 240, 242
Agriscience career, 474–475
Agriscience Fair, 366–367
Agriscience Student Recognition Program, 470–471
Agronomists, 496–497
Air layering, 133–134
Alar (daminozide), 233–234
Allergies to animals, 350
Alligators for food, 283
Almonds, 180–181
American agriculture
  efficiency of farmers, 7
  history of, 25–26
  importance of American agriculture, 4–6
American Civil War, 29
American Farm Bureau, 4–6
Animal science
  animal health, 503
  animal trainers, 502
  careers, 500

cattle reproduction, 501
companion animal care, 501
cooperative extension agents, 501
dog groomers, 502
geneticists, 500
horse industry, 502–503
pet care workers, 501
veterinarians, 503
Animals
  animal health technologists, 503
  animal rights advocates, 378, 387–389
  animal welfare
    Career Development Events, 382–383
    confining animals, 380–383
    defined, 378
    experimentation on animals, 386–387
    management practices, 383–386
    by producers, 379
  cats. *See* Cats
  cattle. *See* Cattle
  companion. *See* Companion animals
  deer. *See* Deer
  dogs. *See* Dogs
  ethical treatment of. *See* Ethical treatment of animals
  as food source, 13
  horses. *See* Horses
  industry. *See* Livestock industry
  for leather, 13
  meats. *See* Meat industry
  for medicines, 13
  mules, 280
  organic agriculture, 258–262
  pork. *See* Pork industry
  poultry. *See* Poultry
  spotted owl, 214

527

support industries, 14
uses for, 268
wild turkeys, 213, 216
for work, 13–14
Annual plants, 145
Anther, 132
Apples, 171, 172–175
Aquaculture
   background, 324–325
   catfish, 325–331
      brood ponds, 327
      characteristics, 326–327
      dissolved oxygen, 329
      fingerlings, 328
      fry, 327–328
      harvesting, 328–330
      roe, 327
      seine, 330
      selective breeding, 327
      warm-water fish, 326
   cool-water fish, 332
   crustaceans, 332–334
      crayfish, 332–333
      defined, 332
      prawns, 334
      shrimp, 333–334
   defined, 324
   ectothermic characteristic of fish, 325
   ornamental fish, 334–335
   Specialty Animal Production Proficiency Award, 330–331
Aquifers, 237–238
Artificial insemination
   biotechnology, 467
   as a career, 501
   for dairy cows, 293

B

B20, 455
Bacteria
   Bacillus thuringiensis (Bt), 93
   in soil, 58
   for vaccines, 471–472
Balance
   flower arrangements, 113
   landscape, 147, 148

BDC (bottom dead center) of engines, 429, 430
Beef industry. *See also* Cattle
   beef statistics, 269
   breed, 269
   calves, 271–272
   carcass, 270
   crossbreeds, 270
   feedlots, 272–274, 501
   mad cow disease, 368
   management practices, 383–385
   purebreds, 270–271
   ration, 272
   stocker operations, 272–273
Bees, 72–73
Bevel square, 403
Biennials, 146
Biodiesel
   B20, 455
   defined, 452
   diesel engines, 453–454
   European and South American consumption of, 455–456
   sources for, 452, 454
   uses for, 454–455
   vegetable oil, 455
Biofuels
   biodiesel, 452–456
   biomass, 456–458
   concerns, 459–460
   defined, 448
   ethanol, 448–452. *See also* Ethanol
   methane, 458–459
Biology, 6
Biomass, 456–458, 460
Biotechnology
   defined, 466
   for food preservation, 466–467
   genetic engineering, 469–474
   reproduction, 467–469
      all-male or all-female embryos, 468–469
      clones, 468
      embryo transfer, 467–468
Birds. *See also* Poultry
   game, 306
   pheasant and quail, 318
   spotted owl, 214

Blanching, 365
Border plants, 257–258
Boring tool, 411
Botanical plant name, 152
Bottom dead center (BDC) of engines, 429, 430
Bovine somatotropin (BST), 291
Brace, 411
Branding of animals, 383–384, 385
Brazilian use of ethanol, 451
Breed, defined, 269
Breeding. See Selective breeding
Broilers
   brooder, 311
   defined, 307
   growing the broilers, 310–314
   hatching the chicks, 308–310
   industry background, 307–308
   processing, 314–315
Bromeliads, 104
Brood ponds, 327
Brooder, 311
Brownback, Sam, 487
BST (bovine somatotropin), 291
Bud scale scars, 66
Bulbs, 100–101
By-product, 450

C

Calves
   for beef, 271–272
   in the dairy industry, 292–293
Canopy of trees, 215, 218
Carbon cycle, 456–457
Carcass of beef, 270
Carcinogen, 233
Career Development Events (CDEs)
   Agricultural Mechanics Career Development Event, 404–405
   agricultural pest education, 88–89
   animal welfare, 382–383
   companion animals, 342–343
   Dairy Cattle Judging and Dairy Foods competitions, 296–297
   FFA International Exchange Program, 10
   FFA Organization, history of, 32–33
   Floriculture Career Development Event, 106–107
   Global Programs, 488–489
   Land Judging Career Development Event, 52–53
   Livestock Judging Event, 276–277
   National FFA Convention, 498–499
   Nursery Operations Proficiency Award, 128–129
   Nursery/Landscape Career Development Event, 154–155
   overview, 489–490
   Poultry Judging Career Development Event, 312–313
   Work Experience Abroad, 11
   World Agriscience Studies, 10–11
Careers in agricultural science
   agribusiness, 496
   animal science, 500–503
   food science, 503–507
   National FFA Convention, 498–499
   natural resources, 507–508
   plant science, 496–500
Carnivorous plants, 104
Carter, Jimmy, 486–487
Carver, George Washington, 200
Castration, 383–385
Catalyst, 454
Catfish production, 325–331
   brood ponds, 327
   characteristics, 326–327
   dissolved oxygen, 329
   fingerlings, 328
   fry, 327–328
   harvesting, 328–330
   roe, 327
   seine, 330
   selective breeding, 327
   warm-water fish, 326
Cats
   cat food, 350–351
   cat scratch fever, 349–350
   for companions, 341, 344
   spaying and neutering, 353–354
Cattle
   cattle drives, 30–31
   cooling of meat, 31

industry. *See* Beef industry; Dairy industry
transportation of, 29–31
CDEs. *See* Career Development Events (CDEs)
Cereal, 192, 371
Chalk line, 407
Cheese, 296–299, 466–467
Chemicals, purpose of, 12
Chewing gum, 11–12
Chickens. *See* Poultry
Chicle, 11
Chisels, 410
Chrysanthemum flowers, 111–113
Citrus fruits, 177–179
Civilizations, 24
Clamps, 414
Clay soil, 54
Climax vegetation, 218
Clone, 468
Clothing
  cotton, 9, 28
  genetic engineering for production efficiency, 473
  linen, 68, 205–206
Coal, 446
Cold frames, 126
Colostrum, 293
Combination square, 404
Combines, 34–35, 193
Combustion, 425
Commodities, 14
Community development in FFA, 240–241
Companion animals
  accessories, 351
  Career Development Events, 342–343
  careers, 14, 501–502
  defined, 340
  diseases and afflictions, 348–350
  ferrets, 340
  health benefits, 342–345
  health care, 351–352
  money spent on, 350
  pet food, 350–351
  responsible ownership, 352–354
  service animals, 345–348
    guide dogs, 345–346
    hearing-ear (signal) dogs, 347
    horses for hippotherapy, 347–348

spaying and neutering, 353–354
types of pets, 341–342
Compass saw, 409
Composting
  manure, 253, 256
  organic fertilizer, 255
Compression ratio, 430
Compression stroke, 429–430, 433
Compression system of engines, 435–436
Confinement operations, 275–277, 380
Cooking oil, 197
  corn, 195
  cotton, 205
  flax, 206
  sunflower, 197
Cooling system of engines, 437–438
Cool-water fish, 332
Cooperative extension agents, 501
Cooperative Extension Service, 37
Coping saw, 409
Corn
  cooking oil, 195
  for ethanol fuel, 8, 195–197, 449–452
  fermentation, 292
  for food, 195
  genetic alteration to produce nitrogen, 472–473
  hybrid, 194
  price increase due to ethanal production, 460
  production, 194
Cosmetics, animal experimentation, 387
Cotton
  background, 202–203
  boll weevil, 200
  for clothing, 9, 28
  cotton engine, 27–28
  cottonseed meal for feed, 205
  growing of, 203
  harvesting, 203–204
  history of cotton production, 26–29
  King Cotton, 28
  processing, 204–205
  Sea Island cotton, 27
  upland cotton, 27, 28
Cover crops, 254–255
Cows. *See* Dairy industry

Crayfish, 332–333
Crop production, 31–35. *See also* Grain crops
  harvesting, 33–35
  sod cultivation, 32–33
Crop rotation, 257
Crossbreed, 270
Crosscut saw, 408, 409
Crustaceans, 332–334
  crayfish, 332–333
  defined, 332
  prawns, 334
  shrimp, 333–334
Cultivars, 173
Cultivators, 12
Customary measuring system, 401–402
Cut-flower production, 104–113
  cut flowers, defined, 104–105
  greenhouse benches, 107–108
  growing cut flowers, 108–111
    lighting, 109–110
    temperature, 110
    timing, 111
  harvesting, 111–112
  physical environment, 105–106
  transportation, 112–113
Cuttings, 133–134
Cycle of engines, 428

D

Dairy industry
  artificial insemination, 293
  BST, 291
  cheese, 296–299
  Dairy Cattle Judging and Dairy Foods competitions, 296–297
  dairy products, 290
  milk production, 291–295
    animal feed, 292
    breeds for, 291
    calves, 292–293
    colostrum, 293
    homogenization, 295
    milk as food, 290
    milk grading, 295
    milk processing, 294–295
    milking, 293–294
    pasteurization, 295
  reproductive technology, 469
  selective breeding, 290
  yogurt, 299
Davis, Jim, 487
DDT, 232–233
Deciduous trees and shrubs, 145
Deer
  for food, 283
  forest habitat, 213
  population control, 388–389
Deere, John, 32–33
Dehorning of animals, 383–384
Deltas, 48–49
Deoxyribonucleic acid (DNA), 469–470
Diesel, Rudolph, 453–454
Diesel engines
  advantages and disadvantages, 435
  background, 453–454
  mechanism, 434–435
  uses for, 434
  vegetable oils for fuel, 8
Digitalis, 9
Disease
  companion animals, 348–350
  to fight weeds, 93
  mad cow disease, 368
  parasites. *See* Parasites
  spread by insects, 84
Dissolved oxygen, 329
Distillation for ethanol production, 449
Distribution of products, 16–17
Division of plants, 134
DNA (deoxyribonucleic acid), 469–470
Docking of animals, 383–384
Dogs
  for companions, 14, 341, 344
  dog food, 350–351
  groomers, 502
  guide dogs, 345–346
  hearing-ear (signal) dogs, 347
  infections from bites, 349
  spaying and neutering, 353–354
  for work, 13–14
Drugs from plants, 9. *See also* Medicine
Drying of food, 362

Ducks, 317–318
Durum wheat, 192

E

E. coli bacteria
   genetic alteration, 471, 473
   in manure as fertilizer, 256
   in processed meat, 369–370
E10, 452
E85, 452
Ecology, 6, 7
Ecosystem
   defined, 218–219
   forest, 218–221
   soil, 57–60
      ecosystem, defined, 57
      microorganisms, 58–60
      plant life, 57
   wetlands, 244–245
Ectothermic, 325
Eggs
   consumption statistics, 306
   hatching, 308–310
   incubators, 310
   layer industry, 315–316
Electronic measurers, 406
Embryo transfer, 467–468
Emphasis. *See* Focal point
Employment for farm-related occupations, 7–8
Endothermic, 325
Energy crisis, 447
Engines
   defined, 425
   diesel. *See* Diesel engines
   internal combustion, 34–35, 422, 426–428
   small. *See* Small engines
   timing, 428
Entomologists, 497
Entrepreneurship SAE, 485
Environment
   community development in FFA, 240–241
   natural resources, 230–231
   nitrate pollution, 242–244
   pesticides, 232–235
   protection of, 230
   renewable resources, 231
   soil erosion, 239–242
   water pollution, 235–239
   wetlands, 244–245
Environmental Protection Agency (EPA), 235, 238, 245
Enzyme, 449
EPA (Environmental Protection Agency), 235, 238, 245
Erosion
   defined, 48–49
   water-deposited soil, 239–242
Ethanol
   from biomass, 458
   from corn, 8, 195–197
   defined, 448
   disadvantages of use, 460
   dry milling, 449, 450
   E10, 452
   E85, 452
   production of, 448–450
   wet milling, 449–451
Ethical treatment of animals
   animal rights advocates, 378, 387–389
   animal welfare, 378–387
Exhaust stroke, 431, 433
Exhaust system of engines, 437
Exotic plants, 104
Experiment stations, 37
Experimentation on animals, 386–387
Experimentation SAE, 484–485
Exploratory SAE, 484
Exports, 14–15

F

Fancy fish, 334–335
Farm factories, 380
Farming, 231. *See also* Agriculture
Farrowing crates, 381
Federal Forest Reserve Act, 221
Feed for animals
   corn, 195
   cottonseed, 205
   flax, 206
   ration, 272

Feedlots, 272–274, 501
Fermentation
　cattle feed, 292
　defined, 449
　for ethanol production, 449
Ferns, 103
Ferrets, 340
Fertilizer
　commercial fertilizers, 253–254
　indoor plants, 162
　lawns, 157–158
　manure, 459
　nitrate pollution, 242–244
　organic agriculture, 252, 253, 254–256
　　composting, 255, 256
　　fish slurry, 255
　　manure, 255–256
　purpose of, 12
FFA Organization
　Agricultural Mechanics Career Development Event, 404–405
　Agriscience Fair, 366–367
　Agriscience Student Recognition Program, 470–471
　animal welfare issues, 382–383
　community development, 240–241
　Dairy Cattle Judging and Dairy Foods competitions, 296–297
　Floriculture Career Development Event, 106–107
　Food for America program, 74–75
　forestry field events, 216–217
　Global Programs, 488–489
　hardware store SAE, 412–413
　high school agricultural education programs, 485–487
　history of, 32–33
　International Exchange Program, 10
　Land Judging Career Development Event, 52–53
　Livestock Judging Event, 276–277
　membership characteristics, 487
　Nursery Operations Proficiency Award, 128–129
　Nursery/Landscape Career Development Event, 154–155
　Poultry Judging Career Development Event, 312–313
　proficiency awards, 198–199
　programs, 487, 489–490
　small engine SAE, 426–427
　Specialty Animal Production Proficiency Award, 330–331
　supervised agricultural experience, 178–179
　Turf and Landscape Management Proficiency Award, 155
　Work Experience Abroad, 11
　World Agriscience Studies, 10–11
Fiber crops
　for cloth production, 202
　cotton, 202–205
　flax, 205–206
Fibrous roots, 75–76
Filament, 132
Fingerlings, 328
Fish industry. See Aquaculture
Fish slurry, 255
Flax, 205–206
Floodplains, 48
Floriculture
　bulb production, 100–101
　cut flowers, growing, 108–111
　　lighting, 109–110
　　temperature, 110
　　timing, 111
　cut-flower production, 104–113
　　cut flowers, defined, 104–105
　　greenhouse benches, 107–108
　　growing cut flowers, 108–111
　　harvesting, 111–112
　　physical environment, 105–106
　　transportation, 112–113
　defined, 100
　FFA Floriculture Career Development Event, 106–107
　florists, 102
　flower arrangements, 113
　international aspect, 100–101
　national aspect, 101
　plants for florists, 102–104

exotic plants, 104
foliage plants, 103
potted plants, 102–103
Flour, 191
Flowers of plants, 71–73
Focal point
flower arrangements, 113
landscape, 147–148
Foliage plants, 103
Food
agricultural methods and environmental protections, 231–232
alligators, 283
careers in food science, 503–507
chickens. *See* Poultry
dairy. *See* Dairy industry
deer, 283
fish. *See* Aquaculture
food chain, 44–45
Food for America program, 74–75
food scientists, 504–505
food technicians, 505–506
genetic engineering for production efficiency, 473
jerky, 362
meat industry. *See* Meat industry
microwave cooking, 361, 365–366
preservation
biotechnology, 466–467
canning, 364
drying, 362
freezing, 365–368
meat, 364
pickling, 362–363
purpose of, 361, 504
reasons for, 360–361
refrigeration, 31, 364–365
salting, 363–364
safety, 368–371
stems, 68–69
Forests
canopy, 215
conservation, 212
ecosystem, 218–221
forestry field events, 216–217

forestry overview, 212
hardwood, 214, 218
harvesting, 212, 213–214, 221–222
industry jobs, 212, 214, 221
natural forest, 214
replanting, 212
softwood, 214
succession, 215, 218, 224
tree farms, 223–224
U.S. Forest Service careers, 508
virgin timber, 213–214, 220
wildlife benefits, 212–213
wood-fiber production, 221–224
Fossil fuels, 446–448
Four-stroke cycle engine
bottom dead center, 429, 430
compression ratio, 430
compression stroke, 429–430
defined, 428
exhaust stroke, 431
intake stroke, 429
operating principles, 428–429
power stroke, 430–431
scavenging, 431
top dead center, 429, 430
two-stroke engine vs., 433–434
Foxglove plant, 9
Framing square, 403
Freeze-drying foods, 367–368
Freezing of food, 365–368
Fruits
apples, 171, 172–175
citrus fruits, 177–179
peaches, 175–176
pome fruits, 171–175
stone fruits, 175–176
supervised agricultural experience, 178–179
transportation, 170–171
Fry, 327–328
Fuel
biofuels. *See* Biofuels
fossil fuels, 446–448
fuel system of engines, 436–437
Fulcrum, 398
Fungi in soil, 59

G

Game birds, 306
Gasohol, 197
Gatherers, 24–25
Gears, 399
Geese, 317–318
Gene mapping, 38, 469–470
Genes, defined, 469
Genetic engineering
    for agricultural pests, 93
    bacteria for vaccines, 471–472
    concerns, 473
    defined, 469
    DNA, 469–470
    monogastric animals, 472
    regulation of, 473–474
    ruminants, 472
Geneticists, 500
Glaciers, 46–47
Global Programs, 488–489
Gluten, 191, 450–451
Gossypol, 205
Grafting, 172, 176
Grain crops
    corn, 194–197. *See also* Corn
    grain, defined, 190
    grasses, 190
    oil crops, 197
    peanuts, 200–201
    soybeans, 197–200
    wheat, 191–193
Grain merchants, 506
Greenhouses
    cold frames, 126
    construction of, 125
    functions, 125
    greenhouse benches, 107–108
    Quonset, 126
    temperature, 110
Groundwater, 237–238
Growing season, 16
Growth stimulants, 254, 258

H

Habitat. *See* Wildlife habitat
Hammers, 411–412
Hand tools
    boring tools, 411
    brace, 411
    care of, 400
    clamps, 414
    cleaning, 401
    customary measuring system, 401–402
    FFA Agricultural Mechanics Career Development Event, 404–405
    hammers, 411–412
    history of, 396–397
    inclined plane, 397–398
    levels, 414–415
    lever, 398
    marking tools, 407
        chalk line, 407
        marking gauge, 407
        pencil or knife, 407
    measuring devices
        defined, 405
        electronic measurers, 406
        measuring tape, 406
        ruler, 405
        zigzag ruler, 406
    metric measuring system, 401–402
    plumb bob, 415
    power tool vs., 397
    safety, 399–400
    saws
        compass saw, 409
        coping saw, 409
        crosscut saw, 408, 409
        holding lumber in place, 409–410
        kerf, 408
        parts of, 408
        ripsaw, 408, 409
        uses for, 408
    screw, 398
    screwdrivers, 413–414
    wedge, 398
    wheel and axle, 398–399
    wood chisels, 410
Hard red spring wheat, 192
Hard red winter wheat, 191–192
Hardware store SAE, 412–413
Hardwood, 214, 218
Harvesting

apples, 174–175
catfish, 328–330
chrysanthemum flowers, 111–112
combine, 34–35, 193
cotton, 203–204
crops, 33–35
equipment, 12
flax, 206
forests, 212, 213–214, 221–222
internal combustion engine, 34–35
nuts, 182–183
peaches, 176
peanuts, 201
problems of, 33–34
threshing machines, 34
wheat, 34, 193
Hatch Act, 37
Herbaceous ornamental plants, 145
Herbicide
defined, 91, 472
nonselective, 91
plant resistance to, 472
selective, 91–92
Heterosis, 224
High school agricultural education programs
Agricultural Education programs, 482
National FFA Organization, 485–487. See also FFA Organization
Supervised Agricultural Experience, 483–485. See also Supervised Agricultural Experience (SAE)
Hippotherapy, 347–348
History of agriculture
American agriculture, 25–26
cattle industry, 29–31
changes in agriculture, 37–38
civilizations, 24
cotton, 26–29
crop production, 31–35
FFA Organization, 32–33
food production, 25
hunters and gatherers, 24–25
scientific research, 35–37
Holly, 134–136
Homogenization, 295

Honeybees, 72–73
Hormones for raising animals, 254, 258
Horses
artificial insemination, 467
grooms, 502
for hippotherapy, 347–348
industry, 280–281
trainers, 503
uses for, 13, 280, 281
Horticulturists, 497–498
Houseplants, 104
Humus, 45, 255
Hunters and gatherers, 24–25
Hybrids
corn, 194
hybrid vigor, 224
trees, 223–224
Hydrological cycle, 235, 236

Ice Ban, 450
Ignition system of engines, 438–439
Iguanas for pets, 341–342
Imports
of agricultural goods, 15–16
defined, 14
Inclined plane, 397–398
Income spent for food, 4
Incubator, 310
Inorganic soils, 46–48
Insects
as agricultural pests, 82–83
biological control, 86
insecticides, 85–86, 260
integrated pest management, 86, 87
life cycle, 86–87
organic pest control, 257–258
parasites, 84. See also Parasites
predator insects, 258
reproduction, 84–85
species, 84
spread of disease, 84
Insulin, 470
Intake stroke, 429, 433
Intake system of engines, 436

Integrated pest management (IPM), 86, 87
Interiorscape, 161–163
Internal combustion engine, 34–35, 422, 426–428
IPM (integrated pest management), 86, 87
Irradiation of meat, 258
Irrigation, 12

J

Jackson, Bo, 487
Jerky, 362
Johnsongrass, 88–89

K

Kerf, 408
King Cotton, 28
Knife or pencil, 407
Koi, 335

L

Lagoons, 243–244, 459
Lambs, 278
Land Grant Act, 36, 37
Land Grant colleges, 36–37
Landscaping
   construction, 153, 156
   defined, 142
   FFA Nursery/Landscape Career Development Event, 154–155
   interiorscaping, 161–163
   landscape architect, 142–143
   landscape architects, 498
   landscape contractor, 143–144
   landscape design, 147–152
      balance, 147, 148
      design principles, 147–150
      emphasis (focal point), 147–148
      landscape plan, 150–152
      planting plan, 152–153
      proportion and scale, 147, 148–149
      rhythm, 147, 149–150
      unity, 147, 150
   landscape designer, 142–143
   landscaping cloth, 130
   lawn maintenance, 157–159
   maintenance, 156–161
      described, 156–157
      lawn care, 157–159
      pruning, 159–161
   plant materials, 144–146
   site inventory and analysis, 146–147
   turfgrass technicians, 500
Lath house, 127, 130
Lawn maintenance, 157–159
Layer hens, 307, 315–316, 381–382
Leaf scars, 66
Leather, 13
Leaves of plants, 69–70
Legumes, 472–473
Levels, 414–415
Lever, 398
Lighting, greenhouse or photoperiodic, 109–110
Lincoln, Abraham, 36
Lind, James, 177
Linen cloth, 68, 205–206
Liners, 120
Linseed oil, 206
Livestock industry
   animals. See Animals
   beef, 269–274. See also Beef industry
   exotic animals, 283
   FFA Livestock Judging Event, 276–277
   horses, 280–281. See also Horses
   meat, 268–269. See also Meat industry
   pork, 274–278. See also Pork industry
   rabbits, 282–283
   sheep, 278–279
   uses for animals, 268
Loam soil, 51, 54
Loess soils, 49
Lubricants, 12
Lubrication system of engines, 439–440
Lumber for buildings and furniture, 9
Lyme disease, 349

M

Mad cow disease, 368
Magneto ignition system, 438

Maintenance, landscape, 156–161
    described, 156–157
    lawn care, 157–159
    pruning, 159–161
Manure
    for biomass, 457–458
    composting, 253, 256
    E. coli bacteria, 256, 369–370
    for electricity, 459
    for fertilizer, 459
    nitrate pollution, 242–244
    organic agriculture, 253, 255–256, 261
Marketing, agricultural, 506
Marking gauge, 407
Marking tools, 407
Mash, 449
McCormick, Cyrus, 34
Measuring devices
    defined, 405
    electronic measurers, 406
    measuring tape, 406
    ruler, 405
    zigzag ruler, 406
Meat industry
    beef, 269–274
    irradiation, 258
    meat inspectors, 505
    meat processing, 505
    overview, 268–269
    per capita consumption, 268
    pork, 274–278
    poultry. *See* Poultry
    rabbits, 282–283
    sheep, 278–279
Mechanical leaf mist area, 131
Media
    artificial ingredients, 121
    bedding plants, 122
    containers, 122–124
    defined, 121
    flats, 122
    soilless media, 121
Medicine
    agricultural products as good nutrition, 6–7
    from animals, 13

drugs from plants, 9
experimentation on animals, 386–387
Methane, 458–459
Metric measuring system, 401–402
Microbiologists, 504
Microorganisms in the soil ecosystem, 58–60
Micropropagation, 134
Microwave cooking, 361, 365–366
Milk expiration date, 371
Milk production, 291–295
    animal feed, 292
    breeds for, 291
    calves, 292–293
    colostrum, 293
    homogenization, 295
    milk as food, 290
    milk grading, 295
    milk processing, 294–295
    milking, 293–294
    pasteurization, 295
Milkfat, 295
Minerals, soil, 45
Mist areas, 130–131
Morrill, Justin, 36
Morrill Act, 36
Motor, 425
Muffler, 437
Mulch, 156
Mules, 280
Mutton, 278

N

National Cattlemen's Association, 385
National Environmental Policy Act, 238
National FFA Convention, 498–499
National FFA Organization, 485–487
National Forest Service, 222
National Pork Producers Council, 385–386
Natural fibers, 202
Natural gas, 458–459
Natural resources, 230–231, 507–508
Natural Resources Conservation Service (NRCS), 242
Natural toxins, 260

Nectar, 72
Nematodes, 59–60
Neutering, 353–354, 385
Nitrate pollution, 242–244
Nitrogen fixation, 254–255
Nodes, 66
Nonpoint sources of pollution, 239
Nonrenewable energy, 447–448
NRCS (Natural Resources Conservation Service), 242
Nursery production
   bedding plants, 122
   Career Development Event, 128–129
   described, 120–121
   FFA Nursery Operations Proficiency Award, 128–129
   growing areas, 124–126
      cold frames, 126
      fields, 124
      greenhouses, 125–126
      outside areas, 127, 130
   growing media, 121–124
   mist areas, 130–131
   plant production, 134–136
   propagation, 131–134
      asexual propagation, 132–133
      cuttings, 133–134
      sexual propagation, 131–132
      tissue culture, 134
Nursing home companion animals, 345
Nutrition of broilers, 307
Nuts
   almonds, 180–181
   defined, 180
   harvesting, 182–183
   pecans, 181–182
   walnuts, 181

O

Oil crops
   cooking oil, 197
      corn, 195
      cotton, 205
      flax, 206
      sunflower, 197
      peanuts, 197
      soybeans, 197
Organic agriculture
   animals, 258–262
   composting, 255
   cover crops, 254–255
   fertilizers, 252, 253, 254–256
      composting, 255, 256
      cover crops, 254–255
      fish slurry, 255
      manure, 255–256
   fish slurry, 255
   insect control, 257–258
   pesticides, 254, 260–261
   production process, 253–254
   rules and regulations, 252–253
   soil amendments, 255
Organic matter, 45
Ornamental fish, 334–335
Ornamental plants
   herbaceous ornamental plants, 145
   popularity of, 10–11
Osmosis, 364
Ovary, 71, 131
Overstory of trees, 215, 218
Ownership SAE, 485

P

Paper, 9
Parasites
   in confinement operations, 381, 383
   defined, 84
   toxoplasmosis, 348
Pasteurization, 295
Peaches, 175–176
Peanuts, 197, 200–201
Pears, 171
Pecans, 181–182
Pencil or knife, 407
Per capital consumption of meat, 268
Pesticides
   Alar (daminozide), 233–234
   carcinogen, 233
   DDT, 232–233
   defined, 232

food safety, 368
organic agriculture, 254, 260–261
purpose of, 234
Pests
defined, 82
FFA education, 88–89
genetic engineering, 93
indoor plants, 163
insects, 82–87
integrated pest management, 86, 87
lawn care, 157
organic pest control, 257–258
weeds, 88–93
Petroleum
for cloth, 202
fossil fuels, 446–448
Pets. *See* Companion animals
pH of soil, 54
Pheasant, 318
Phloem, 67–68
Photoperiodic plants, 109–110
Photosynthesis, 70
Pickling of food, 362–363
Pine trees, 218
Pistil, 71, 131
Pitcher plant, 104
Placement SAE, 485
Plane, 397–398
Plans
landscape, 150–152
planting plan, 152–153
planting schedule, 152–153
Plant nurseries, 120
Plant science
agronomists, 496–497
defined, 496
entomologists, 497
horticulturists, 497–498
landscape architects, 498
turfgrass technicians, 500
Plantations, 224
Plants
for clothing, 9, 28, 68, 205–206
for food, 8–9
Food for America program, 74–75
functions, 66

growing of, 8–12
for lumber, 9
medicines, 9
ornamental plants, 10–11
paper, 9
in soil ecosystem, 57–58
structure
flowers, 71–73
leaves, 69–70
roots, 75–76
seeds, 73–74
stems, 66–69
Plow for sod cultivation, 32–33
Plumb bob, 415
Plummet, 415
Point sources of pollution, 238
Pollination
bees, 72–73
flowers, 71–73
fruit trees, 173–174
process of, 132
Pollution
nitrate, 242–244
sediment, 240
water, 238–239
Pome fruits
apples, 171, 172–175
defined, 171
pears, 171
Pork industry, 274–278
breeds, 274–275
confinement operation, 275–277
farrowing, 275–276, 381
finishing, 275, 277–278
genetic engineering, 472
management practices, 385–386
Pots for plants, 122–124
Potted plants, 102–103
Poultry
broilers
brooder, 311
defined, 307
growing the broilers, 310–314
hatching the chicks, 308–310
industry background, 307–308
processing, 314–315

540

Career Development Event, 312–313
consumption of, 306
ducks, 317–318
eggs, 306, 308–310, 315–316
game birds, 306
geese, 317–318
layer hens, 307, 315–316
pheasant, 318
quail, 318
reproductive technology, 469
selective breeding, 307
turkey production, 316–317
Power stroke, 430–431, 433
Power tools, 397
Prawns, 334
Predator insects, 258
Preservation of food
  biotechnology, 466–467
  canning, 364
  drying, 362
  freezing, 365–368
  meat, 364
  pickling, 362–363
  purpose of, 361, 504
  reasons for, 360–361
  refrigeration, 364–365
  salting, 363–364
Processing of agricultural products, 17, 38
Proficiency awards
  basis of award, 490
  Fruit and Vegetable Production Proficiency Award, 178–179
  Nursery Operations Proficiency Award, 128–129
  overview, 198–199
  row crops, 198–199
  Soil and Water Management Proficiency, 52–53
  Specialty Animal Production Proficiency Award, 330–331
  Turf and Landscape Management Proficiency Award, 155
Propagation. *See also* Reproduction
  air layering, 133–134
  asexual propagation, 132–133
  cuttings, 133–134
  defined, 131
  division, 134
  grafting, 172, 176
  micropropagation, 134
  pollination, 132
  sexual propagation, 131–132
  tissue culture, 134
Proportion and scale
  flower arrangements, 113
  landscape, 147, 148–149
Protozoa in soil, 59
Pruning, 159–161
Psittacosis, 349
Pulleys, 399
Purebred cattle, 270–271

Q

Quail, 318

R

Rabbit production, 282–283
Rabies, 348
Railroads for cattle transport, 29–31
Range managers, 508
Ration, 272
Refrigeration, 31, 364–365
Renewable energy, 456
Renewable resources, 231
Reproduction. *See also* Propagation
  artificial insemination
    biotechnology, 467
    as a career, 501
    for dairy cows, 293
  technology, 467–469
    all-male or all-female embryos, 468–469
    clones, 468
    embryo transfer, 467–468
Reptiles for pets, 341–342
Research on animals, 386–387
Research SAE, 484–485

Resources
　　natural, 230–231, 507–508
　　Natural Resources Conservation
　　　　Service, 242
　　renewable, 231
Retail store manager, 506–507
Rhizobia, 58
Rhythm, landscape, 147, 149–150
Ringworm, 349
Ripping a board, 409
Ripsaw, 408, 409
Rocky Mountain spotted fever, 349
Roe, 327
Roots of plants, 75–76
Roundworm, 348
Row crops
　　defined, 190
　　fiber crops
　　　　for cloth production, 202
　　　　cotton, 202–205
　　　　flax, 205–206
　　grain crops
　　　　corn, 194–197. *See also* Corn
　　　　grain, defined, 190
　　　　grasses, 190
　　　　oil crops, 197
　　　　peanuts, 200–201
　　　　soybeans, 197–200
　　　　wheat, 191–193
　　Supervised Agricultural Experience,
　　　　198–199
Rulers, 405–406
Ruminants, 472

S

SAE. *See* Supervised Agricultural
　　Experience (SAE)
Salting of food, 363–364
Sauerkraut, 363
Saws
　　compass saw, 409
　　coping saw, 409
　　crosscut saw, 408, 409
　　holding lumber in place, 409–410
　　kerf, 408

　　parts of, 408
　　ripsaw, 408, 409
　　uses for, 408
Scale
　　flower arrangements, 113
　　landscape, 147, 148–149
Scavenging, 431
Science, defined, 6
Scientific research
　　defined, 35
　　experiment stations, 37
　　history of, 35–36
　　Land Grant colleges, 36–37
Screw, 398
Screwdrivers, 413–414
SCS (Soil Conservation Service), 242
Sea Island cotton, 27
Seeds of plants, 73–74
Selective breeding
　　catfish, 327
　　dairy cattle, 290
　　poultry, 307
Service animals, 345–348
　　guide dogs, 345–346
　　hearing-ear (signal) dogs, 347
　　horses for hippotherapy, 347–348
Shade cloth, 127
Sheep industry, 278–279
Shrimp, 333–334
Site inventory and analysis, 146–147
Slash and burn land clearing, 262
Slavery, 29
Small engines
　　adaptability, 422
　　care and maintenance, 424
　　combustion, 425
　　diesel, 8, 434–435. *See also* Diesel
　　　　engines
　　four-stroke cycle engine
　　　　bottom dead center, 429, 430
　　　　compression ratio, 430
　　　　compression stroke, 429–430
　　　　defined, 428
　　　　exhaust stroke, 431
　　　　intake stroke, 429
　　　　operating principles, 428–429

power stroke, 430–431
scavenging, 431
top dead center, 429, 430
two-stroke engine vs., 433–434
internal combustion engine, 34–35, 422, 426–428
motor, 425
operating principles
  cycle, 428
  four-stroke cycle engine, 428
  generally, 424–425
  power production, 425
  stroke, 428
  two-stroke cycle engine, 428
SAE, 426–427
safety, 423–424
systems
  compression system, 435–436
  cooling system, 437–438
  exhaust system, 437
  fuel system, 436–437
  generally, 435
  ignition system, 438–439
  intake system, 436
  lubrication system, 439–440
  starting system, 440
timing, 428
two-stroke cycle engine
  compression stroke, 433
  defined, 428
  exhaust strokes, 433
  four-stroke engine vs., 433–434
  intake strokes, 433
  operating principles, 428, 431–433
  power stroke, 433
Smith-Hughes Act, 37
Smith-Lever Act, 37
Soft red winter wheat, 192
Softwood, 214
Soil
  amendments, 255
  bacteria, 58
  clay, 54
  defined, 44
  ecosystem, 57–60
  erosion, 48–49, 239–242
  fungi, 59
  humus, 45
  inorganic soil, 46–48
  Land Judging Career Development Event, 52–53
  loam, 51, 54
  loess, 49
  microorganisms, 58–60
  minerals, 45
  organic matter, 45
  pH, 54
  plant life, 57–58
  protozoa, 59
  purpose of, 44–45
  soil amendments, 255
  Soil and Water Management Proficiency, 52–53
  Soil Conservation Service, 242
  soil deposited by wind, 49–50
  soil horizons, 55–56
  soil profile, 55
  soil scientists, 508
  texture, 51, 54
  water-deposited soil, 48–49
  wind-deposited soil, 49–50
Sows. *See* Pork industry
Soybeans, 197–200
Spaying, 353–354
Speed square, 404
Spotted owl, 214
Squares
  bevel square, 403
  combination square, 404
  defined, 402
  framing square, 403
  speed square, 404
  try square, 403
Stamen, 71, 132
Starting system of engines, 440
Statistics
  efficiency of American farmers, 7
  importance of American agriculture, 4–6
Stems of plants, 66–69
Stigma, 71, 131, 132
Stillage, 450

Stock plants, 120
Stone fruits, 175–176
Stroke of engines, 428
Style, 71, 131
Succession of forests, 215, 218, 224
Sugarcane for ethanol production, 451
Supervised Agricultural Experience (SAE)
    Agricultural Education, 483
    background, 483–484
    Dairy Cattle Judging and Dairy Foods competitions, 296
    exploratory, 484
    Fruit and Vegetable Production Proficiency Award, 178–179
    hardware store, 412–413
    ownership/entrepreneurship, 485
    placement, 485
    proficiency awards, 198–199
    research/experimentation, 484–485
    row crop proficiency awards, 198–199
    small engines, 426–427
    Specialty Animal Production Proficiency Award, 330–331

T

Tang chisel, 410
Taproots, 76
TDC (top dead center) of engines, 429, 430
Technology. *See* Biotechnology
Temperature, greenhouse, 110
Terminal tip, 66
Texture of soil, 51, 54
Ticks, 349
Timed system of mist area, 130–131
Timing, greenhouse plants, 111
Tissue culture, 134
Tools. *See* Hand tools
Top dead center (TDC) of engines, 429, 430
Topiary, 160
Toxoplasmosis, 348
Tractors, 12
Trade balance, 14
Transportation of products
    cattle, 29–31
    distribution network, 17
    flowers, 112–113
    fruits, 170–171
Tree farms, 223–224
Trees for landscapes, 145
Tropical fish, 335
Try square, 403
Turf and Landscape Management Proficiency Award, 155
Turfgrass technicians, 500
Turkey production, 316–317
Two-stroke cycle engine
    compression stroke, 433
    defined, 428
    exhaust strokes, 433
    four-stroke engine vs., 433–434
    intake strokes, 433
    operating principles, 428, 431–433
    power stroke, 433

U

United States Department of Agriculture (USDA)
    establishment of, 36
    food safety, 369
    genetic engineering guidelines, 474
    irradiation of meat, 258
    organic agriculture, 252–253
    rabbit meat nutrition, 282–283
Unity
    flower arrangements, 113
    landscape, 147, 150
Upland cotton, 27, 28
U.S. Forest Service careers, 508
USDA. *See* United States Department of Agriculture (USDA)

V

Vaccines, 471–472
Variety and unity, 147. *See also* Unity
Vegetable oil for fuel, 455
Venus flytrap, 104
Veterinarians, 351–352, 503
Vitamin C, 177
Vocational Agriculture, 482–483
Volcanoes, 50

W

Walnuts, 181
WASS (World Agriscience Studies), 10–11
Water
    aquifers, 237–238
    forest ecosystem, 219
    groundwater, 237–238
    hydrological cycle, 235, 236
    nitrate pollution, 242–244
    nonpoint sources of pollution, 239
    point sources of pollution, 238
    pollution, 235–239
    rainfall, 237
    water-deposited soil, 48–49
WEA (Work Experience Abroad), 11
Wedge, 398
Weeds
    Bacillus thuringiensis (Bt), 93
    biological control, 92
    controlling, 90–93
    defined, 88
    diseases to fight weeds, 93
    herbicide, 91–92
    Johnsongrass, 88–89
    as pests, 89–90
Wetlands, 244–245
Wheat, 191–193
Wheel and axle, 398–399
White wheat, 192
Whitney, Eli, 27–28
Wholesalers, 506
Wild turkeys, 213, 316
Wilderness areas, 222
Wildlife habitat
    deer, 213
    forest ecosystem, 220
    in forests, 212–213
    more land for organic agriculture, 262
    spotted owl, 214
    wild turkey, 213
Wind, soil deposited by, 49–50
Wine, 466–467
Winter wheat, 192
Wood chisels, 410
Wood for biomass, 457
Wood-fiber production, 221–224
Woody landscape plants, 145
Wool production, 279
Work Experience Abroad (WEA), 11
World Agriscience Studies (WASS), 10–11

X

Xylem, 67–68

Y

Yogurt, 299

Z

Zigzag ruler, 406
Zoonoses, 348